SMARTER MANUFACTURING

顶级办公设计集成
Top-class Office Integration

智造

本书编委会 \ 编

中国林业出版社
China Forestry Publishing House

相忘于江湖

陈宪淳（2001~2011年 Mi2 创意总监 / 2012年 anySCALE任督设计 创始合伙人）

每一本书都应该有自己存在的意义，那么，这本书存在的意义在哪？

编辑向我约稿的时候应该是半年前的事情了，我记得他们打过几次电话给我，因为感觉比较陌生，我当时并没有答应供稿，只是说让他们把以往的样刊发给我看看，然后再考虑，在接下来相当长的一段时间，我陆陆续续收到了杂志社的一些原始的排版小样，期间也接到很多的沟通电话，探讨了对这本书的种种看法和建议，这让我开始觉得他们是在用心做一件事情，虽然青涩，但可贵，也没细细考虑，直觉引导我最后将自己最近做的准备留给另外一本发行量很大的知名杂志的两个案例投给了他们，然后他们又看似轻描淡写，实质变本加厉地问我能否为这本书写个序，得寸进尺的家伙。

在写序言前，同样道理，我要求将排好版的书本内容发给我看看，文件收到后我让公司的设计师们仔细阅读，然后提自己的想法，自己则断断续续花了一些时间前前后后翻阅了几遍，书本的文字说明我基本上没有读，因为是图册类的书籍，我觉得图片才是最真实有效的诉求，有时候配图文字的表达反而束缚了图片本身因为浏览者自己的不同阅历而产生的难能可贵的不同理解，毕竟，读书的过程，本来就是读者与书内容之间产生独特的感应才有意义。

设计行业现在的发展异常迅速，迅速得让很多的设计师都来不及"停顿"（还不是"沉淀"）就发现自己已经过时，人云亦云，疲于奔命。商业设计的强烈需求一方面让设计蓬勃振奋，同时又将设计的灵魂扼杀在疲惫不堪中，草图、平面图、效果图、施工图都应该是设计，假如不这么匆忙的话。任何一个有前途的行业都需要一个萌芽、发展、壮大、混乱、洗牌、回归、精致的过程，现在我们所处的阶段就是"混乱"阶段，技术派在不停抄袭行业内顶尖设计师的最新手法，非常高效但简单粗暴地"拿"；思想派另辟奇径却大多数闭门造车，想法独特之余难免无病呻吟想"给"。"拿"和"给"之间的平衡关系其实就是这个行业有序发展的一个关键所在，设计师谋生很简单，但脱离了谋生的需求后应该怎样去给自己定位？我们到底应该怎么去做？

设计本身是一件很快乐的事情，我们热爱我们的工作，以至于我们把设计当成自己的事业，而不单单是养家糊口的行当，设计师应该是激情飞扬而又含蓄饱满的，我们其实是在实现自己的梦想，我们可以很个性，甚至不需要团结，但必须为这个自己倾情的行业贴砖加瓦，尽自己的能力去提高整个行业的标准，这个时候，最简单而又最需要的是放慢自己的忙乱不堪的脚步，静下心来，整理一下自己的手头的项目，问问自己，我们是在敷衍么？客户对设计的要求也许可以是粗糙的无所谓的,但设计师对设计的执着却永远必须是细腻的偏执的。

因为这样,我们必须不停地去充电,技术层面上的加强很好操作,做加法即可,紧跟Autodesk,Adobe,Google,Corel,Apple等行业先锋的脚步;思想层面上的加强其实更多的是做减法,必须尽量去清空自己固有的思维定律,重新拾回一颗愉悦而敏锐的心,如小孩般见到什么都是清新的。成长让我们付出最大的代价是见怪不怪,我们或许没有办法返老回童,那就只能尽量去忘却自己,回归童趣。每每见到公司的小孩随手在白板画出的他们认为的所谓空间的规划图,让人目瞪口呆,惊讶不止。我们寻寻觅觅了好久的老师,其实一直就在我们的身边,鲜活跳跃着。

我认识一个从业多年的设计师,有一次让我去参观他的工作室,整个空间中满满地陈列了这么多年来他做的一些作品的手工模型和手稿,大大小小的,各种材质都有,他先是绘声绘色地对着这些如珠如宝的模型讲自己的辉煌经历,接着意味深长地说出他想将这些项目出书的愿望,最后他充满疑惑地问我一个问题,为什么最近很长的一段时间他觉得停滞不前,而且客户越来越少了?我笑了笑,没有直接回答,然后在临别的时候只是淡淡地回了一句:假如有一天你心甘情愿地把你的这一堆东西全扔进垃圾桶,你就有救了。

编辑曾经问我我的作品不是放在前面有没有问题,当然没有问题,我们大部分时间会把自己看得太过重要,其实根本没有人会这么热烈地去在乎你。这本书的作品来自各个不同的城市区域,区域带给作品的性格是模糊的,但模糊背后隐藏着的特点却也是清晰的,我们可以通过这本画册大致去了解中国目前设计市场的整体水准,这和武侠小说的各个门派的感觉有点类似,区域市场的成熟度很多程度上决定了作品本身的气场强弱,即使是用一种设计手法,工艺及认知水平的差异也会让作品呈现出几乎完全不一样的氛围,即使是似曾相识,也依然有可能个性鲜明。这点很好理解,大师们都或多或少有自己的惯用手法,但往往能巧妙融合到每一个新的作品中却能让你浑然不觉,风格正是靠这种巧妙的重复积累出来的。在这本书看到的每一个手法的应用,不管多么熟悉,都有可以成为你独有的风格元素,你可以很随便地翻这本书,假如有一天无意中能看到一些恒定不变的东西的时候,属于你的时代就来了。

所以,对于设计师来说,这本画册存在的意义就在于针对不同的案例或细或粗"看"完(不是"读"完),然后忘却,该吃吃,该喝喝,睡醒后重新回到设计的江湖中,拔剑出鞘,凌风不动,呆若木鸡,无招无式,心自飞扬。

Known from the Arena

Chen Xianchun (General Creative Director from 2001 to 2011/Partner and Designer of anySCALE in 2012)

Each book shall have its own meaning. Then, what's the meaning of this book?

It has passed half a year since the editor asked me to write a book preface. I remember that they have called me for several times. As there was some strange for me, I did not agree with them. I told them that they can send me some former samples and I would consider about it. During the following long period, I have received many original typesetting samples from publisher. I also have received many calls on discussing the opinions and suggestions on this book, which gave me a feeling that they were doing one thing with heart, not mature but very precious. Without any further consideration, my intuition leaded me to send my prepared two cases, which I was going to leave to the other well-famous magazine publisher to them. After it, they asked me that whether I can write a preface for this book even. What a greed request!

Before writing this preface, for same reason, I asked them to send me the prepared book. When receiving the documents, I invite our company's designers to read them carefully and then propose their own opinion. For me, I spend time to read it time by time. The words description is not my consideration, because it is an atlas book. In my opinion, the picture is the most real and effective information. Sometimes, the words even limit the different understanding about the picture for their different life experience. After all, the reading's value is lying in the unique resonance between the reader and book.

The design industry has been developed in a rapid way, which gives many designers a feeling that they have found them oldish already without even one second "stop". To meeting people's requirements, they have tired for living. The strong requirements of business design have allowed the design industry developing rapidly and also have killed designers' soul by the exhausted work. If it is not that busy, the draft drawings, plans, renderings, construction drawings should be design works. Any promising industry need a process of sprout, developing, growing, confusion, shuffling, turn to delicate. At present, we are staying at a "confusion" phase. The technology group are copying the top designers' latest techniques and "taking" their skills in a very rude and stupid way. The ideological group always takes another way by isolating with the outside world. Therefore, their ideas are always strange but moan to "give". The balance relationship between "giving" and "taking" is the key points to decide whether this industry can be developed in an orderly way or not. Living on design is easy, but what should we do to position ourselves after addressing the living problem.

Design is a happy thing itself. We love our work so that we like to take design as our career but not only to support family. Designers shall be full of passion. In fact, we are pursuing to realize our own dream. We can be individual, even need not be unity. However, we must give our passion and talent to improve the standards for this industry. At this time, the simplest and most demanding is to slow our step and calm our heart to sort up the existing project. We shall ask ourselves "are we in a perfunctory?" The customer requirements on design may simple, but the designers are always pursuit for perfect and delicate forever.

Just because of it, we must charge ourselves continuously. It is easy to improve the technical part by keeping with the advanced technology like Autodesk, Adobe, Google, Corel, Apple and so on. However, for ideology part, it is better to reduce the existing ideas. We shall try to empty our fixed methods but revive our pleasant and smart heart, like a naive child. The biggest price for grow is indifference. Perhaps we have no way to go back to our childish time. Then, we can try our best to forget ourselves and return the children's happy time.

Each time, when I see the children drawing on the paper what they want to draw, they all give me a big surprise. The teacher is at our side all time but we even have looked for him for such a long time.

I met an experienced designer. Once a time, I visited his office. In the limited whole space, there were laying all his manual model and manuscripts of years, big or small with all kinds of materials. Firstly, he stood in front of these cherished models and told us his brilliant experience vividly. Then, he said that he wanted to put these projects into book. Finally, he asked ma a question that why he felt stagnant for a long period and the customers were fewer and fewer. I smiled to him without answer. Finally, when I was leaving, I told him that if you could drop all your existing models, you would find the way.

The editor has ever asked me that whether the works have any problems. Of course not, most of our time is focusing on ourselves too mush. In fact, nobody will care you too much. This works of this book come from different cities and areas. The regional character brought to the works is vague, but their feature is clear. We could understand the overall standard of Chinese design market through this book. It is similar to the martial arts in different classes. The maturity of the regional market has determined the works popularity. Even adopting a same design technique, the difference of technique and understanding will affect the works and show a totally different sense. Even they are similar, they are still distinctive. It is easy to understanding. The masters always have their own treating methods and they can merge them into each new work without any traces. The style is formed by repeating these techniques. Each technique in this book, no matter how familiar with, can be part of your own elements. You could read it easily. If one day you find some permanent things, then it's your time. Therefore, to designers, the meaning of this book is focusing on different cases. You can watch (not read) them carefully or easily. Then, go back to your life, after waking, when you return your design work, you will feel easily and freely to charge the design by your heart.

目录 CONTENTS

008 | 北京放射
Radial in Beijing

016 | Dunmai 办公室
Dunmai Office

026 | 东方IC创意办公室
Imagine China Creative Office

032 | 2K Games China 仟游软件科技（上海）有限公司办公室
Office of 2K Games China (Shanghai)

040 | Concrete 之家
House of Concrete

048 | 优吧办公室
Urban Bar Office

054 | and…, SuperPress, Superbla 办公室
and…, SuperPress, Superbla Office

062 | YS 工作室
YS Office

068 | 支付宝上海有限公司办公室
Office of Alipay.com Co., Ltd. (Shanghai)

074 | 北京物质性
MATERIALITY in Beijing

084 | 佳敏企业办公室
Office of Carbing Enterprise

090 | 隐巷设计顾问有限公司办公室
Office of XYI Design Co., Ltd.

096 | Groupm 群邑北京办公室
Office of Groupm Beijing

102 | 道和设计机构办公室
Office of Daohe Design Institution

108 | 东灿五金贸易有限公司全球总部
Headquarter of Dongcan Hardware Trading Limited Company

116 | 瑞士苏黎世谷歌 EMEA 工程中心
Google's New EMEA Engineering Hub in Zurich, Switzerland

124 | 汉诺森设计机构办公室
Office of Hallucinate Design

128 | 红坊办公室
Red Town Office

134 | 红桃网办公室
Office of Aceona.com

138 | 思维繁殖场
Idea Breeding Ground

146 | 经典国际设计机构（亚洲）有限公司办公室
Office of Classic International Design Agency (Asia) Co., Ltd

154 | 某平面设计公司办公室
Office of a Planar Design Company

158 | 雷迪有限公司办公室
Office of Leidi Limited

170	深圳派尚设计公司办公室 Office Space of Shenzhen Panshine Interior Design Co., Ltd.
176	盘古投资办公室 Pangu Investment Office
184	上海创盟国际建筑设计有限公司新办公室 New Office Space of Archi-Union Architecture Design INC. (Shanghai)
192	答案之门 The Way Out
200	MARYLING服装中国区总部 MARYLING Clothes China Office
206	腾讯科技(第三极)办公楼 Office Building of Tencent Technology (the Third Pole)
212	玲珑 Ling Long
216	厦门共想设计办公室 Office of Gongxiang Design
222	朗诺经贸办公室 Office of Langnuo Economic and Trading
228	"亚邑"办公室 Office of "YAYI" Design
234	杭州意内雅办公室 Yineiya Space Design for Offices in Hangzhou
242	赢家商务中心 Beijing Winner International Business Service Co., Ltd.
250	树下 Under the Tree
258	山二集团企业总部 YAMANI Group
266	设计部办公室 Ministry of Design Office
274	Mediacom北京办公室 Mediacom Office in Beijing
280	飞利浦创新科技园LED中心 LED Center of Philips Innovation Science and Technology Park
290	宽银幕 大家庭 Wide Screen, Big Family
296	Vital Déco办公室 Vital Déco Office
300	艾迪尔新办公楼 Ideal New Office Building
308	真工设计黎明事务所办公室 Office of Z-WORK Design Associate
316	艺谷(北京)公司新办公室 The New Beijing Office Eegoo

北京放射
Radial in Beijing

设计单位：SAKO 建筑设计工社
照明设计：Masahide Kakudate Lighting Architect & Associates, Inc.
设 计 师：迫庆一郎、青山周平、今福佑一郎、巨亮
项目地点：中国北京市
建筑面积：3100 m²
主要材料：MDF 板、彩色玻璃
摄 影 师：Misae HIROMATSU—锐景 Photo

二十七层平面布置图

二十九层平面布置图

2020年的广告公司将会是彩虹——个性与协作的办公空间。从公共通道延伸至窗边的彩虹吊顶，让你在和公共通道保持距离的同时，享受开放的办公空间。顶部颜色以代表CIG的红色为中心，呈环状向两边渐变。会议室和办公室大量使用的彩色玻璃，与顶部的彩虹色保持整体感，室内办公家具与整体装饰风格保持统一。彩虹其丰富的色彩表示员工多样的个性，环状又表示个性的集合。

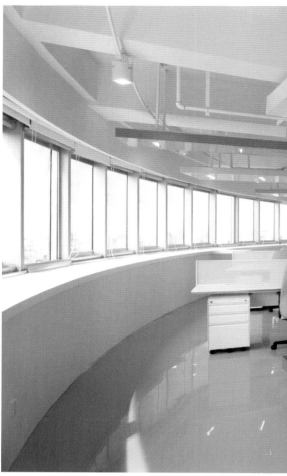

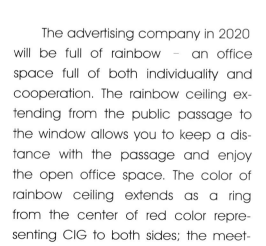

The advertising company in 2020 will be full of rainbow – an office space full of both individuality and cooperation. The rainbow ceiling extending from the public passage to the window allows you to keep a distance with the passage and enjoy the open office space. The color of rainbow ceiling extends as a ring from the center of red color representing CIG to both sides; the meeting room and the office are full of painted glass which maintains integrity with the rainbow color of the ceiling; office furniture in the room is in color of a ring rainbow in accordance with the overall decorative style and the rich colors show a variety of employees' personalities and the ring shape shows the collection of personalities.

Smarter manufacturing —— Top-class office integration

Smarter manufacturing —— Top-class office integration

013

Smarter manufacturing —— Top-class office integration

Dunmai 办公室
Dunmai Office

设计单位：**Dariel Studio**
设 计 师：**Thomas Dariel**
项目地点：中国上海黄浦区中山南路**1029**号幸福码头**3**号楼
主要材料：真石漆、大理石、木地板、马赛克
项目负责：侯胤杰
摄 影 师：**Derryck Menere**

Dunmai办公室坐落于南外滩的幸福码头时尚创意园区，其前身为上海幸福摩托车厂，园区在车水马龙的南浦大桥地块并不张扬，华灯初上的时候，柠黄色的灯光映照在古朴裙楼间，绵绵绿竹也映出婆娑的灯影，勾勒出一处静谧、悠闲、开阔的创意空间。南外滩是上海的标志，也是上海人从上个世纪延续至今不灭的情节，老码头，外马路，十六铺……单是看到这些地名，土生土长的上海人，脑海中就会缓缓飘起黄浦江的涛声，氤氲着晨雾，响起巨轮和轮渡汽笛的鸣咽。如今的南外滩，不仅延续了老外滩的古雅奢华和董家渡的历史繁荣，也连接代表着科技、未来的外滩一体化工程和世博园区。

这个项目的客户是来自澳门的创意活动主办公司，在充分考虑了客户的特性和要求后，设计师 Thomas DARIEL 决定将其打造成符合客户定位的充满活力的现代创意办公室，同时带给员工一种愉悦、轻松的工作氛围，并且利用高科技的产品来体现办公室的实用性和功能性。

内部建筑结构的重塑

由于之前这座四层的老厂房内部结构的限制，因此设计师决定只保持外立面的历史感，而打破室内所有原有结构，重塑一个具有线条感的开放性三层楼结构。

这种开放性的建筑结构一改原先大楼结构的弊端，使得办公室的采光更加明亮并且加大了空间感。为了体现客户开放性的要求，在重塑结构时，设计师特别将中庭打造成一个挑空开放的区域，不论位于哪一楼，都能方便的与每一楼层的同事沟通，并且采用长型白色钢琴漆办公桌，即满足了友好的工作氛围需求，又便于员工之间的交流。

"在公园里工作—在办公室玩乐"

室内环境的设计中，此次办公室的设计主题是"在公园里工作—在办公室玩乐"。

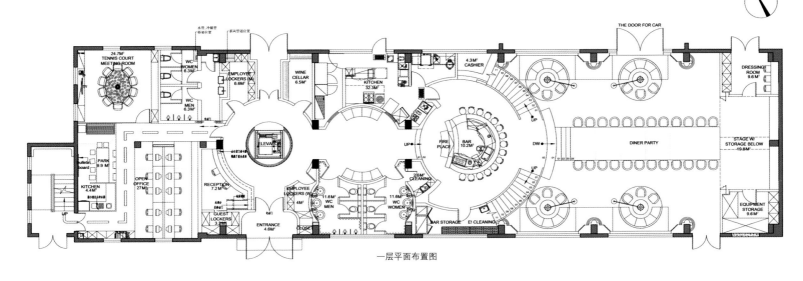

一层平面布置图

这个主题被不遗余力地运用于整个空间结构中，以展现出一个先锋和有趣的办公室空间。

"在公园里工作"

首先，运用灯光设计将整个空间点亮得如同室外一样明亮。在配色上运用纯白，将这个原本阴暗陈旧的老厂房改造成一个通透明亮、简洁纯净的办公区域。在我们的作品中经常能见到的强烈色彩，这次也毫不例外，采用粉色或嫩绿来点缀空间，形成对比以期给人处于自然的感觉。

其次，在重塑内部空间结构时，以大树枝干为灵感，营造出生动的线条感，并且在墙面上运用向上排列的方格抽屉，让人联想到一格格花盆，种植在里面的植物不断延伸向上，更好地体现了在"在公园里工作"的主题。

为了使整个办公空间都能让人有置身公园的感觉，设计师尽可能地移入了户外公园的概念元素：铺设在桌椅下的绿色草坪好似在草地上办公；墙上的园艺工具装饰；带有童年色彩的秋千随意地摆放在一旁，工作累了还能回归下童年；桌子上两边特意留出放置盆栽的花道，使整个办公室到处都春意盎然……

"在办公室玩乐"

热衷于在室内设计中展示另人出乎意料的设施，制造错位摆放和幽默感的惊喜一贯是 Dariel Studio 的设计风格。

此次设计的主题就是为了增强客户的公司活力，并使传统、严谨的办公室变得乐趣横生。

在这里，工作也似乎变得有趣轻松起来：办公室的地面上马路交错，会议室的设计以网球场为雏形：网格的玻璃隔墙，蓝色的网球场塑胶地面，嵌着网球的桌子……

休息室的入口被特别设计成看似电梯的门。当你远远看到一部电梯，殊不知正想上楼的瞬间，却是一扇自动门通向同样具有创意的卫生间。

就连最细节处的卫生间的墙面，也借鉴了法国著名艺术家的涂鸦风格，将小时候的游戏机中的形象作为马赛克的图案，在任何一个地方都不遗余力地想要呈现愉悦的感觉。 在体现时尚和轻松的办公环境氛围的同时，设计师并没有忘记办公室所应具有的功能和实用性，通过运用高科技，在很多处运用玻璃的隔墙和门，既创造了一个开放透明的空间，又更方便员工之间的交流。

怀着对历史建筑的崇敬之情，设计师完好地保留了建筑的外立面，并巧妙地改变内部空间的结构，不仅继承了老厂房的工业气质，还通过创意布局和设计使其变成时髦的办公之所，符合本案客户的角色定位和需求。

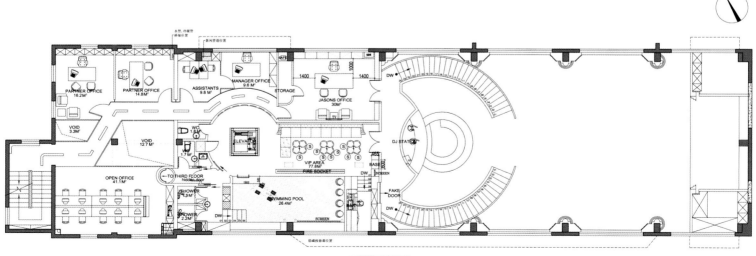

二层平面布置图

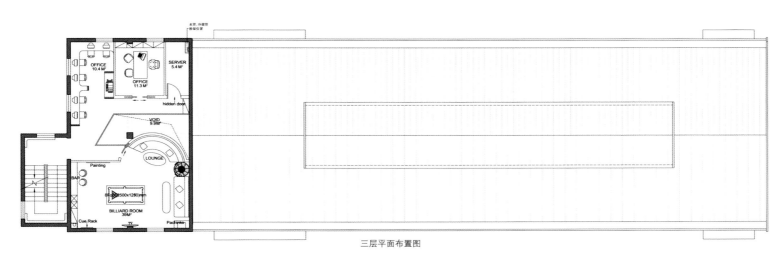

三层平面布置图

This project is an office space nested in an old motorcycle factory. It is located in the creative plaza on the South of the Bund, a place that serves as a remembrance to the familiar stories about old Shanghai. The Shanghainese old docks, the tide of the Huangpu River, and the whistles from ships in the morning also acknowledge the nostalgic feeling of the city's ancient past. This location is not only the continuation of the old Bund's elegance and historical prosperity, but also serves as a connection to the newly developed business districts.

After considering all the requests and specificities of the client, a company from Macao organizing creative events, Thomas Dariel decided to construct the office into a pleasurable, relaxing and modern place that reflects the company's dynamism and creativity as well as serving their professional needs. This kind of atmosphere fits to a creative space as well as being functional by using high-tech products.

An internal architectural reconstruction

To memorialize this old 4-floor factory building, the creator kept the building's historical fa?ade, but completely transformed the internal structure into an open, high 3-floor volume arrangement.

This architectural approach opened up the space to obtain brighter and wider rooms, meanwhile revealing the original structure of the building. The internal reconstruction also conveys a friendly atmosphere to the open space required by the client. A 3-floor high central patio, furnished with large white lacquer desks, was built so that colleagues can easily communicate with each other despite what floor level they are on.

"Work in the Park – Play in the Office"

In terms of interior design and atmosphere, the concept developed in Dunmai is "Work in the Park – Play in the Office."

This concept is reflected in the overall structure and in every detail of this edgy and humorous-looking space.

"Work in the Park"

First, the entering light itself has been designed to pervade the whole space, so that one could feel a sense of being outside. The color scheme consists of pure white in order to change the office's previous image of a dark and old factory

into a clean and simple display. As often in our projects, bold colors such as vivid pink or green are used ubiquitously for contrast and for developing natural energy.

Furthermore, the shape of the new internal structure is inspired by the branches of a tree. Drawers on the wall let people imagine that all the plants growing in that space are spreading along the wall and up to the ceiling.

In order to make people feel as if they were surrounded by nature in the given space, the designer attempted to move all the elements of a garden to inside the office: grass lawns under chairs and tables, gardener's tools designed on the walls, swings displayed during afternoon breaks that exhibit the sweet memory of childhood, and a groove for flowerpots on tables.

"Play in the Office"

Playing with interiors by displaying unexpected facilities, misusing objects or creating jokes is part of the Dariel Studio's signature.

In this project, the concept is used to enhance the vitality of the company and to offer a unique and humorous work environment. Dunmai office fully subverts the appearance of a traditional strict office: there are crossroads lining the floors; a tennis court style meeting room with a grid glass wall; floors modeled after blue tennis grounds; and a table set featuring tennis balls.

The restrooms' entrances are designed to look like open elevator doors. Thus, when one is in search for an elevator, he or she will be surprised to find that it is actually a bathroom.

Even the toilet walls are creatively designed, for the designer pays homage to a famous French artist street style by using images from video games to decorate tiled mosaics. The design illustrates that working in an office can be a joyful and unique experience. An office space can be open and transparent, just like the glass walls and doors of many individual spaces in Dunmai Office.

The new design is a reverence to the previous historical building Dunmai Office once was, as well as to the strong heritage and the trademark qualities of the industry's construction style. This creative indoor arrangement makes this space a definite trendy and innovative office, and, at the same time, answers the client's identity and satisfies his requirements.

Smarter manufacturing —— Top-class office integration

东方IC创意办公室
Imagine China Creative Office

设计单位：**Dariel Studio**
设 计 师：**Thomas Dariel**
项目地点：上海
建筑面积：**700 m²**

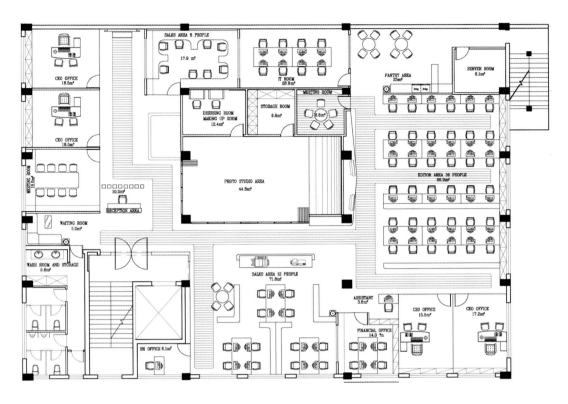

　　本项目位于静安区一幢老厂房的5层，靠近静安寺。静安寺是一所古老的佛家寺院，位于本区域的中心地带。东边紧邻黄浦区，北面是苏州河，静安区是中心区之一，也是上海人口最为密集的区之一。但这幢老厂房周围却是相当安静并有宽阔的林荫道。

　　对于本案，我们充分考虑了客户的特性。东方IC是一家专业的多元化视觉资源提供商和图片技术服务商，服务于国内和国际市场。为了体现他们与在中国各地摄影师联系的庞大网络，其新办公室需要很好地服务于其专业需求并且要反映出其活力和创意。

　　结合客户的专业领域和需求，我们将已有空间重新设计，使用天然材料，打造成纯净明亮的LOFT创意空间，根据各个独立的专业区域，重新布局了各个实用功能空间，使销售及市场、编辑、摄影、人事、IT及管理部门在视觉上归于各个不同的区域。

　　7 m高玻璃屋顶的摄影棚位于中心区域，象征着公司的核心业务，它指引着整个空间并组织引领着办公室里其他的工作区域。我们认为拍摄和编辑图片是一项需要安静环境的工作，因此我们采用了天然材料并混合了水泥和木头，运用明亮的白色并结合了风水创造出了具有"禅"的意境的小花园。

　　开放式工作区配备了大型白色钢琴漆办公桌，满足了友好的工作氛围并使员工交流起来更轻松。配色方案也仍然大面积选用纯净的白色。而会议室的玻璃门选用了红色以平衡并打破白色的单调。同样，在图书馆区，几组暗色调几何型书架延续了强烈的画面感和视觉元素。

　　我们非常小心没有破坏其空间的原本结构，继承了老厂房的工业气质，通过创意布局使其变成时髦的办公室，符合东方IC视觉资源提供商的角色定位。

The project is located on the fifth floor of a former factory building in Jing'an District, named after Jing'an temple, an ancient traditional Chinese Buddhist temple located in the heart of the area. Boarding the Huangpu District in the east and Suzhou creek to the north, Jing'an is now one of the central districts and one of Shanghai's most densely populated neighborhoods. Yet the surrounding of this old building is relatively peaceful and benefits of a wide tree lined street.

For this project, we had to consider the specificity of the client, Imaginechina, which is a leading photo agency in China, producing and syndicating features and photos for both of the international and local markets. Representing the largest network of Chinese photographers located in all regions and provinces of China, their new office have thus to serve their professional requirements as well as reflect their dynamism and creativity.

Combining the field of the client and its needs, we redesigned the existing into a shape-pure, natural materials, bright loft-type creative space doted of strong practical space functions, corresponding to distinct professional fields. Indeed, the design, if homogeneous, makes a visual differentiation between Sales and Marketing, Editing, Shooting, HR, IT and Management areas.

The shooting studio with its seven meters ceiling and its glass roof is central located. Standing for the core business of the company, it

Smarter manufacturing —— *Top-class office integration*

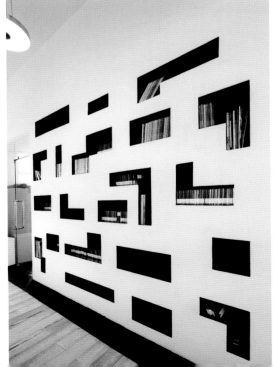

brings light to the entire space and organizes the rest of the office that gravitates around it. We thought that shooting and editing were tasks requiring a peaceful environment and thus designed accordingly by preferring natural materials with a mix of concrete and wood, bright white color and adding a fengshui touch while creating a "zen" garden.

The open working space is furnished with large white lacquer desks and answers the requirement of friendly atmosphere and easy communication between the staff. The color scheme again uses mainly pure white. Yet a bright red color applied on the meeting rooms doors engenders contrast and balance while breaking the all-white rhythm. Likewise the libraries, several groups of dark geometric open bookcases that constitute strong graphic and visual elements.

If an utmost care was given not to touch the original space structure, strong heritage and trademark of the industry building style, the creative arrangement makes this space a definitely trendy office, accordingly to Imaginechina's positioning --- visual resource supplier for both home and overseas.

2K Games China 仟游软件科技（上海）有限公司办公室
Office of 2K Games China (Shanghai)

设计单位：上海明合文吉建筑设计有限公司
设 计 师：徐明、Virginie Moriette
项目地点：上海市兆丰大厦
建筑面积：**1800 m²**
主要材料：钢板、水泥板、玻璃、多层板、地坪漆
设计时间：**2010**

Smarter manufacturing —— *Top-class office integration*

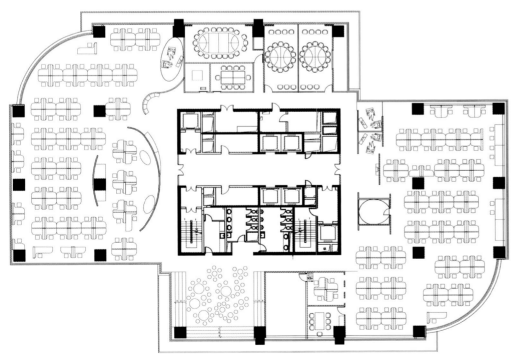

2K Games China 仟游软件科技（上海）有限公司是一家成立于2006年初的游戏开发与发行公司，是美国Take-Two/2K公司的全资子公司。随着中国业务的不断发展，原本由明合文吉设计的办公室I期、II期都已容纳不下不断壮大的办公队伍，因此公司租下了原本办公楼上的整个楼层，也就有个如今的III期项目。

III期项目仍然由明合文吉负责设计，整个回环型的楼面经过合理规整，有序地将行政区、办公区、会议室、茶歇区各个功能区域容纳其中。办公室的前台接待区与行政区融为一体，那里有一道非常醒目的墙面，它特有开豁流畅的弧线构造，赋予了原本四方规整的空间更多动感。一侧的等候区域由直立的白色PVC管围拢出一个圆柱状空间，配上特

033

为其设计的绿色沙发椅,成了极具观赏性的小场景。

Ⅲ期的办公区域仍然沿用红白相间的色调规划,然而灯光系统进行了重新设计,全新的灯具不仅造型玲珑纤巧,其散发的柔和光线不会在电脑屏幕上产生光斑,更加便于员工使用。办公区域内随处可见的玻璃书写板,总拖着一个逗号状的尾巴,仿佛方格漫画中出场人物的对话框,非常生动有趣。此外,办公区域的隔断被做成了城市大楼的白色剪影,其上还配备镂空的小窗,透过各个小窗,所看到的办公环境将是另外一种风情。

如上所述的体验式情景式设计在会议室中被运用到极致。设计师将篮球、排球、乒乓球、拳击台等元素糅合到空间的设计上,变幻出与之相应的会议区域:篮球室内的桌面好似球场中的部分区域,侧边的墙面上更安置了一个真实的篮球架;乒乓桌成了乒乓会议室的会议桌;排球室内,白色的球灯被安置成排球击出时的弧形动线,很是形象;红色的沙发加上护栏,一个拳击台式的会议室就这样构成了。这样的设计不仅符合游戏软件公司的特定身份,也让会议产生了趣味性,很有特色。此外因为公司的一款007游戏,设计师特意找来一个硕大的"集装箱",外面漆成亮丽的红色,内里则是酷酷的黑色,添加一个狙击手的剪影造型,007会议室就这样诞生了。

员工的茶歇区内不仅放置了众多的桌椅,还配置了不少用于娱乐放松的器具,最受欢迎的当数斯诺克台球桌,心情烦闷时,约几个同事来这里上一句,彼此交流,说不定还会有意想不到的创意产生。侧边的区域还放置了阶梯型的条凳,据说这在定期举办的员工桌球赛时能聚集不少观众。

Established in early 2006 and wholly-owned by American Take-Two/ 2K, 2K Games China (Shanghai) is engaged in development and issuance of games. With the development of business in China, offices in phase I and phase II, which are designed by Design MVW, cannot hold the gradually increased personnel, therefore, the Company has rented the whole floor above the offices, which is known as the phase-III project.

The phase-III project is stilled designed by Design MVW. After being designed properly, the whole winding floor will hold in order the administrative area, office area, meeting room and rest room. In the office reception area integrated as a whole with the administrative area, there is a striking wall which makes the square and regular space felt more dynamic owing to its specially open and smooth camber structure. In the waiting area at the side, the cylindrical space surrounded by white upright PVC pipes, together with the green sofa designed for it in particular, makes up a small ornamental scene.

In spite of the reapplication of red and white colors, the phase-III office area is designed to be equipped with a new lighting system where the new exquisite and dainty lamps and lanterns will be easier for employees to use owing to their soft light which will not cause spots on the computer screen. The glass writing board can be seen everywhere in the office ar-

ea. It is interesting with its comma-shape tail looking like a dialog box of the character present in comic strips. What's more, the partition in the office area is of white building silhouette and is installed with small hollow windows through which the office will be seen in an environment with a different feeling.

The aforementioned design based on experience and scene is applied to an extreme in the meeting room. The designer applies basketball, volleyball, table tennis, boxing ring and other elements in the space design, resulting in corresponding meeting areas: The basketball room is set with a table looking like certain part of the basketball court and with a real basketball stand on the side wall; the table tennis room is equipped with a table tennis table; the volleyball room contains white globe lanterns which are arranged into a camber line like being formed after volleyball is hit, showing vividness; the boxing ring room is composed of red sofa with rails. Such design is suitable for the Company's specific identity as a game software maker and makes meetings interesting, showing unique features. In addition, based on the Company's game called 007, the designer designs a 007 room, with a large "container" painted into bright red outside and into cool black inside and with a sniper silhouette.

The rest room for employees is equipped with a lot of desks and chairs as well as entertaining facilities among which snooker tables are most popular. In case of depressed mood, several colleagues can get together here to communicate with each other, and an idea may be born unexpectedly. In the side area, ladder benches are arranged. It is said that many viewers are attracted here when the employee snooker match is held regularly.

Smarter manufacturing —— Top-class office integration

Concrete 之家
House of Concrete

设计单位：concrete
设 计 师：rob wagemans、ulrike lehner、sofie ruytenberg、erik van dillen、volkert vos、erikjan vermeulen and lisa hassanzadeh
室内木工：roord binnenbouw
钢 结 构：smederij van rijn
电　　力：kronenburg团队
项目地点：荷兰
建筑面积：480 m²
摄 影 师：ewout huibers

concrete之家的建筑为Oudezijds Voorburgwal 101的附属建筑,历史可追溯至1880年。在新的空间里,concrete团队的35名专业人员分布于建筑的4个楼层,阁楼用作材料库。

在这里,建筑师、视觉设计师、室内设计师及平面设计师以跨学科的方式参与项目设计工作。每一办公间均铺设未经处理的木地板,墙壁刷为白色,配备两张办公桌,办公空间面积各不相同。每位员工配备一台铝制小推车,用于储存个人物品、图纸、文具和文件夹。小推车在上部加垫,并附有橙色的垫层,以对小型的小组会议创造额外的座位。

每一办公桌上方悬挂两盏"蜘蛛"吊灯,使整个墙壁具有同一性。concrete之家要求每个团队成员收集10张可激发灵感的图片,现已收集400张图片,分别张贴于四面墙上。最后,将与现有项目相关的文件保存在黑色文件夹和A3硬纸箱,然后储存在档案柜中。书写板与磁条提供了展示现有项目和灵感的空间。

中央楼梯/图书馆

中央楼梯是团队成员之间的物理连接,而灵感(我们的藏书)正来源于此。将图书馆置于楼梯中间,加上二楼的咖啡室,走廊就从一个流通空间变成了公

共空间。

更好的房间——正式会议室

为容纳众多的群体、客户和相关人员召开会议，将一楼的"最佳"房间设为正式会议室。此会议室配备大型的黑色橡树会议桌，镶嵌黑色皮革；窗前放一黑色陈列柜，存放比例模型、励志书籍、奖品和用于报告的电视。会议桌上方悬挂"蜘蛛"吊灯，周围放置12把荷兰工业老椅，用灯芯绒做坐垫，墙上悬挂艺术品。

最好的房间——非正式会议室（名为snug）

Snug为非正式会议室，位于三楼，不对客户开放。此空间用于举行设计会议，配有桌子或舒适的长椅沙发，周围摆放着杂志。

厨房

房位于地下室后方，正对庭院，是此项目中的重要元素。此空间充满怀古气息，包括柜台和火炉设施，可以在此烹制每日的午餐，每月还要举行一次烹饪竞赛。转盘餐桌周围放置10把埃姆斯椅，颜色各异。大理石地面和墙壁铺贴原产的白色荷兰"witjes"瓷砖。

为增设额外席位，在走廊对面添设一件经典家具：午餐桌。座位由定制的之字形钢制成，配备的钢制座椅与桌子融为一体并用橙色混凝土粉末喷涂。

模型+巨型比例模型工作空间

比例模型工作空间位于地下室前部，与建筑入口相邻。concrete创建比例模型，将其用于设计初期的交流，以达到有效调整空间视觉性、三维性、灯光和材料的目的。

The HOUSE OF CONCRETE is based in a listed building on the Oudezijds Achterburgwal in the middle of the red light district. The building is an annex of the Oudezijds Voorburgwal 101 and dates back to 1880.

The entire team of 35 professional people is distributed over the 4 storeys of the building and the attic is dedicated to the materials library.

At this new location architects, visual marketeers, interior designers and graphic designers work on the projects in an interdisciplinary way. Concrete builds brands, produces the architectural and interior designs

and urban development plans, in combination with the main presentations and scale models.

the work rooms

Four works room are divided over three floors (1st, 2nd and 3rd floor). Every work room has an untreated wooden floor, the walls are painted white and have two worktables with a variable number of work spaces. Every member has its own aluminium trolley, to store personal belongings, drawings, stationary and binders. The trolley has a cushioned top, with concrete-orange upholstery to create additional seating for small team meetings.

Above every worktable hang two 'dear ingo' chandeliers and one entire wall is immersed in the concrete identity; every team member was asked to collect 10 inspirational images and now there is a collection of 400 frames divided over 4 walls. Finally, archive cabinets provide storage for all current projects filed in black binders and cardboard A3 boxes. Pin-up boards and magnetic strips provide space to exhibit current projects and inspiration.

the central staircase / library

The listed building provides many additional square meters in the hallway and staircase which could have a double function. As the central staircase is the physical connection between the team members, inspiration (our books) is the binding metaphor in our work. By placing the library in the middle of the staircase in combination with a coffee pantry on the 2nd floor, the hallway transforms from a circulation space into a social space.

the better room - formal meeting

room

To accommodate meetings with large groups, clients and relations there's a formal meeting space at the 1st floor in the monumental 'best' room. The room is furnished with a large black oak meeting table with black leather inlay, a black display cabinet placed in front of the windows for scale models, inspirational books, awards and a TV for presentations. A 'dear ingo' chandelier hangs above the table, 12 gispen chairs, upholstered in black corduroy surrounds the meeting table and art pieces hang on the wall.

the best room - informal meeting room (the snug)

The snug is an informal meeting room on the 3rd floor where officially no clients are allowed. It's a space to have design meetings at the table or on comfortable chesterfield couches surrounded by our collection of magazines.

the kitchen

The kitchen in the back of the basement, looking over the courtyard, is a crucial element in the HOUSE OF CONCRETE. The antique space, including counters and a stove facilitates daily lunches together and once a month a cook-off. A 'lazy susan' table surrounded by 10 eames chairs in various colours. Restored marble flooring and the walls are covered with original white tiles called 'witjes' in dutch.

To create additional seating, again the corridor offered sqm to put in an iconic piece of furniture: the lunch tunnel. A zigzag custom-made steel seating with integrated table which is powder coated concrete-orange.

Models + Monsters scale-model workplace

The scale-model workplace is in the front of the basement, next to the entrance of the house. Concrete creates scale-models to communicate the design in the preliminary phase, which is the most efficient method to accommodate the visualisation of spaces, 3d environments, light and materials.

优吧办公室
Urban Bar Office

设计单位：Naco Architectures/纳索建筑设计事务所
(www.naco.net)
设 计 师：Marcelo Joulia
项目地点：上海市乌鲁木齐路
建筑面积：**140 m²**
主要材料：不锈钢、金刚板、防古砖、玻璃钢
摄 影 师：徐文磊

Smarter manufacturing —— Top-class office integration

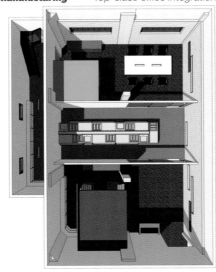

优吧是一家专业的调酒师培训机构，提供私人或职业调酒师培训，研发新口味的鸡尾酒，以及承接各类活动的调酒服务。业主希望自己的办公空间能反映出"Urban 优吧"充满活力的品牌精神，同时在功能上也需要满足他们特殊的使用需求。我们将"玻璃房"培训室放在整个空间的中央，用其隔开吧台和办公区域。与此同时，"玻璃房"错层的设计也使得空间的动线更加有层次感，让员工进出办公室时体验"穿越"的乐趣。

Smarter manufacturing —— *Top-class office integration*

Urban Bar is a professional agency engaged in private or occupational bartender training, researching and developing cocktail with new flavors, and providing bartending service for various activities. The owner hopes that the office space can reflect the energetic brand spirit of "Urban" and satisfy special utilization demands for functions. Therefore, we arrange the "glass" training room in the center of space to separate bar counter and office area. Meanwhile, the staggered layer design of "glass" training room makes the dynamic lines of space feel more layering and offers employees the fun of "crossing" upon their passing in and out of the office.

and⋯, SuperPress, Superbla 办公室
and⋯, SuperPress, Superbla Office

设计单位：Naco Architectures／纳索建筑设计事务所
　　　　　（www.naco.net）
设 计 师：Marcelo Joulia
项目地点：上海市八号桥一期
建筑面积：80 m²
摄 影 师：徐文磊
撰　　文：Amber Chan

"Galaxy",对应的中文意象是:银河,星系,以及——一群杰出的人。初识 and…, SuperPress, Superbla 的人可能不能立刻了解这几家公司之间的关系,其实只要记得它们是一个"Galaxy"就好。

整个办公空间由纳索(Naco)建筑设计工作室设计完成。"白色"以"统领者"姿态霸道占领这个空间,让人想到北欧的皑皑白雪,茫茫无际状。所有颜色光线聚集在一起,最后集大成的便是——白,最单纯不过也最博大深广的颜色。纳索上海设计总监 Margaux Lhermitte 女士告诉我们纯白创造出解除禁锢感的空间,原初纯朴未经加工,像是一块为不设限的创意准备的空白画布,用"a-n-d…"的CEO——Ruth Ang 的话来说,这里就像是"裁缝的剪裁工作室",激发灵感任人挥洒。

点缀其间的红色尤为醒目。红色圆形大地毯铺成第一层办公空间的圆心,环绕其周的鲜红椅垫,还有攀着旋转楼梯蜿蜒上楼的红地毯,像是贯穿整个空间的血脉。红色是拥有磅礴力量的色彩,热情热忱是它的专属语言,令人充满活力,将沉闷呆滞一扫而净。

"红白相间"是 Naco 和这个 galaxy 钟爱的色彩搭配,SuperPress 的红白 logo 就是其中例证之一。

在色彩上还占有一席之地的是绿和紫。这两色的选择也不是临时起意,其灵感来自以极简美学著称的时尚设计师 Jil Sander 的时装发布会,它们也都拥有旺盛的能量,浓艳鲜妍。很难在这里的"绿"和"紫"前面加上描述的定语,很难用"墨绿"、"翠绿"、"草绿"、"青绿"或是"茎紫"、"晶紫"、"红紫"、"浅紫"来定义它们,因为它们的颜色层次随着时光流转在人们不经意间悄悄变幻。"因为创意一直都在变。"Margaux 说,"那颜色为什么不?"

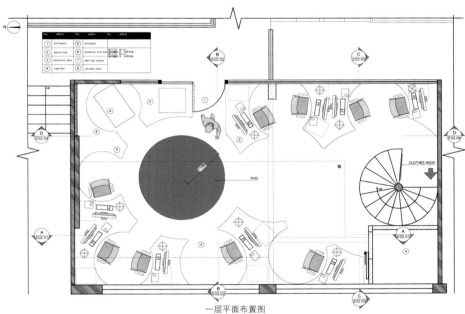

一层平面布置图

最让格子间总是抱怨苦闷的白领们眼前一亮的,或许是这里造型各异独特新奇的办公桌。虽说造型各异,但也并非无迹可循,它们像是原本完整的一大块实木,用半径不同的圆模挖空,也像是裁缝剪过圆形布块后剩下的边角料。这些"边角料状"的办公桌在 Naco 的鬼斧神工下,华丽转身为一个特别的存在——"让在里面工作的人既可以轻松自在相互谈话",随时随项目变化团队方阵,又能让每个人拥有自己独享的专属空间。打破约束和隔阂,是纳索为这个空间设计自始至终的坚持点。就是"有机感(Organic feel)",这是 Ruth 用的词。

Smarter manufacturing —— Top-class office integration

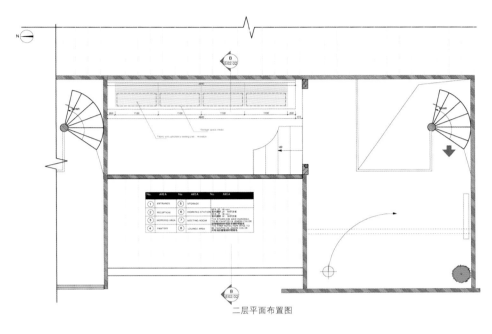

二层平面布置图

旋转楼梯是Ruth的心水所在。在设计上,这个旋转楼梯贯通上下空间,打破平面的单调,创造出另一个新的工作氛围。它不仅在外形上以灵巧蜿蜒大得分,最特别的地方是——楼梯下通常被废弃的空间被镂空为一个"我们的骄傲"展示台,不用来放包放杂物,有资格登上这些展示台的都是值得让他们感到骄傲的作品。

"审美意识"在这里有着非凡地位。小到台灯,纸篓,大到挂壁艺术品,都经过精心挑选,因为它们都承载"激发灵感"的重任。不得不提的是旋转楼梯上空悬的白色大鸟笼,常令访客感到惊奇。"这正好就是我们希望达到的效果:惊讶——惊奇——惊喜。"Margaux笑谈,"因为这是为未来而设计的办公空间。"它并不需要按照既定的标准框架来打造,鸟笼为这里营造出更丰富的情绪,类似于孩童时代的嬉戏情怀。这个鸟笼中你看不到被捆缚的鸟儿:因为创意不可以也不应该被"关禁闭"。

但是,在这一个完全开放式的空间,私密性要如何解决?

其实这也是有意为之。纳索和在这个空间工作的团队都不希望设置任何封闭或隐蔽空间,因为"头脑风暴"是他们的每日必修课,他们本来就应该持续在团队中工作,相互间以及和客户间交换各自的想法和创意。

身为建筑设计师,在被问到最钟爱这个空间的哪个部份时,Margaux说:"是它透过透明门呈现给过客们的样子。"这一个纯白却也不乏温馨感的空间一直吸引人们驻足观赏,他们会取出相机为它留影,他们交头接耳希望了解更多。对于一个建筑设计师来说最高的成就莫过于——启发人们思考,提问。

参观后更了解这个空间以及在里面工作的团队SuperPress,Superbla,a-n-d···——他们正在开展的是全新的事业,像是一个喷溅创意,和媒体以及新科技相连的星系。它给人们呈现的是一个有关美好有关未来的愿景,令人为之而激动。

"Galaxy" means Milky River and galaxies, as well as —outstanding people. The people who acquaintance with SuperPress, Superbla may not understand the relationship between them. You only need to remember "Galaxy".

This office is designed and completed by Naso Construction Design Studio. "The white color" is covered this space in a "dominate" way, letting people imagine of the endless snow in Nordic. All the colors and lights focusing here have formed a single performance—white, which is the most purity and greatness color. Ms. Margaux Lhermitte, the Design Director of Naso in Shanghai, told us that the white has created the most freedom space. This pure color looks like a bland canvas with endless creative. In CEO—Ruth Ang's word, it is just as a "tailor studio" to inspire people's soul.

The red color is dotted in the space. The round and red carpet has paved the first floor's space, surrounding with the bright red cushions and the red carpet on the spiring stairs, which looks like the veins of the entire space. Red is a majestic force color with enthusiasm. Enthusiasm is its exclusive language, letting people fill with vitality and let the boring go.

"The red and white color" is the favorite color adopted by Naso and galaxy. The red and white logo of SuperPress is one example.

The other important colors are green and purple. These two colors are selected not by moment. Its inspiration comes from the fashion designer Jil Sander's Fashion Collection. They are possessing the genetic power, brightly and colorful. It is hard to add description words before "green" and "purple". It is hard to add "ink green", "bright green", "grass green", "light green" or "dark purple", "crystal purple", "red purple" or "light purple" to define them, because their color are changing slightly with times. "Because the creation is always changing", Margaux said, then, why not col-

or?

"The Best Grid" is always surprised by the boring whiter colors, maybe because there are different shapes and unique strange office table. Thought they are different shapes, they have certain rules. They are looking like an original complete wood and be hollowed with different diameters. They are also looking like the rest parts of a piece of cloth cut by tailor. These "cast-off" shape office tables are special and unique through great skills of Naso—"letting the workers can have a easily communication". These tables can be changed into different style in accordance with the different projects and let each people possess their own special space at the same time. Breaking through the constrains and separation is a great point insisted by Naso for this space. That is "Organic feel", which is described by Ruth.

The spiring stairs is the focus of Ruth. On the design, this spiring stairs passes through the whole space between ups and downs, which has broke the boring plain and created a new working atmosphere. It is not only unique in terms of the flexible out shape, the most great point is that the under part of stairs is always

Smarter manufacturing — *Top-class office integration*

abandoned and the designer hollowed it as a "Proud Exhibition" platform. It is not used to store the bags and packages. The products that can be displayed on are all products we proud of.

"Aesthetic consciousness" has a major position here. From the lamp, wastebasket to the wall-hanging art, they are all selected carefully, because they are all shouldered the responsibility of "inspiration". The one I have to mention is the hanging white birdcage on the spiring stairs, which brings visitors surprise often. "It is just the effect what we want to express: surprise—amazing—comfortable" said Margaux, "because it is an office designed for future." It doesn't need to shaped in accordance with the traditional standards. The birdcage here is to create more fun, which is looking like the childish sense. In this birdcage, there is no bird, because the innovation can not and shall not be "closed".

However, in this completed open space, how to reserve a private space?

In fact, it is also set by purpose. Naso and this work team are all not hope to set any closed or secrecy space, because "brain storm" is the necessary class for them each day. They of course shall work for this team and exchange their own ideas and innovations with customers and themselves.

As an architecture designer, when asked which part of the space is my favorite, Margaux said "the most favorite one is the transparent door's appearance shown to visitors." This white but warm space are attracting visitors and makes them take photos with it. They talk with each other for more information. To an architecture designer, the most achievement shall be this—to inspire people to ask and meditate.

After visiting, we know more about this space and the work team SuperPress, Superbla, a-n-d...they are developing the brand new career, looking like a spray innovation, connecting with the media and new technology. The scenery it shown to people is a beautiful landscape about future, which let people exciting.

YS工作室
YS Office

设计单位：汤建松·丰造设计顾问有限公司
设 计 师：汤建松
参与设计：张雅书、吴志江
项目地点：厦门市吕岭路长安大厦601室
建筑面积：**120 m²**
摄 影 师：刘腾飞

我们用简明的框架构图，将如宣的白墙描绘成封面，在清澈的明镜里，交互的刨花隔断菱角张扬，而轻柔的白沙纷纷淡淡。只有那传世的青花，隐在边缘里，自顾自美丽……

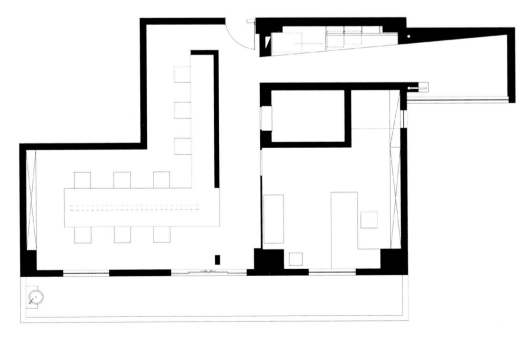

Simple framing composition is applied to draw the white paper-like wall into a cover where flowers fly among water caltrops in the clear mirror-like water while white soft sands lie glimmeringly, which is like a song, that is, only the blue and white colors on porcelains handed down from ancient times are hidden in the edges, focusing on their own beauty…

支付宝上海有限公司办公室
Office of Alipay.com Co., Ltd. (Shanghai)

设计单位：上海思域室内设计工程有限公司
设 计 师：宋毅
参与设计：王南钢
项目地点：上海市证大五道口广场
建筑面积：**4500 m²**
主要材料：生态木、**PVC**、**Inteface**块毯、定制金属隔断、挂板
撰　　文：宋毅

二十八层平面布置图

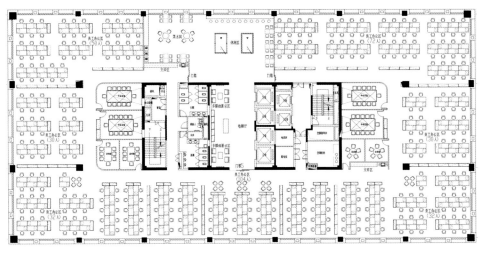

二十九层平面布置图

作为一个差不多"地球人都知道"的知名网络平台,近期常常也是经济、网络、媒体的焦点和宠儿,公司员工平均年龄26-27岁,卧虎藏龙,达人、潮人不少,创意、策划点子颇多。这也给了设计师不小压力。

支付宝近年扩张势头凶猛,导致办公空间相对拥挤,人员较密集。设计的初衷就是在"密集中寻找空间的呼吸感"。门厅地面一改往昔常规的室内装饰材料,采用生态防腐木,给人户外轻松之感,结合S型可调节角度金属百叶围合的空间形成连续、流畅、舒缓的空间,同时提供了阅览、招聘、私语的多样性活泼空间。整体空间呈开放性,强调沟通、交流、自由。在有限的环境中,相对集中辟出公共活动空间与工作区衔接对话,无论工作、咖啡、游戏,这里都是舞台,任你自由扮演角色。色彩、空间造型和家具的鲜活让年轻的人心中天天都是阳光灿烂的日子。

As a network platform "known by everybody on the earth", Alipay.com Co., Ltd. recently becomes the focus and favorite in terms of economy, network and media. In the company, employees share an average age from 26 to 27 and there are undiscovered talents, many intelligent persons and trendsetters as well as lots of originalities and ideas. All these put enormous pressure on the designer.

In recent years, the Company is expanding rapidly, leading to crowd office space and dense employees. Therefore, the design intends to make employees "breathe in dense space". The hall floor is made of ecological anticorrosive wood instead of common materials for interior decoration, to make people relax. With the space enclosed by S-shape metal shutters whose angles are adjustable, a continuous, smooth and comfortable space is formed, which can be applied as a diverse area for reading, employing and talking. The whole space is open and emphasizes communication, exchanging and freedom. In the limited area, a public area is opened up to connect with the working area. Here is your stage and you can play your roles freely whatever you do – work, have coffee or play games. The color, space shape and fresh furniture all aim to make the young people feel sunny in their heart everyday.

Smarter manufacturing —— *Top-class office integration*

北京物质性
MATERIALITY
in Beijing

设计单位：**SAKO** 建筑设计工社
照明设计：**Masahide Kakudate Lighting Architect & Associates,Inc.**
设 计 师：迫庆一郎
项目地点：北京市
建筑面积：**110 000 m²**
主要材料：自行设计铝板、大理石
摄 影 师：**Misae HIROMATSU**—锐景 Photo

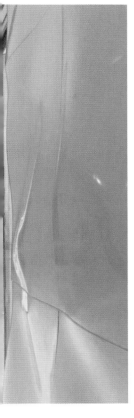

我们设计的"新三里屯"项目的南侧,规划有建筑面积为46万m²的"三里屯SOHO"的城市开发项目。我们负责其办公部分的内装设计。首先,我们按设计的整体方针进行了样板间的设计。

样板间的墙壁上无规则地挂着一些细长的、不锈钢材质的吊柜,通过这些吊柜即可把样板间的这种纵深形状强调出来,又能给空间增添一种动感。另外,吊柜的内侧贴着4种颜色的皮质材料,时尚而又华丽。使用这种颜色,是为了和三里屯整体街道的气氛相吻合。

隔断采用可扭曲投射的透明亚克力棒,柔和委婉地把办公区域分割开来。工作台与外墙壁的线条相吻合,呈曲线型。

我们提倡的是在具有领导潮流的街区,在自由访问式办公系统(non-territorial)里的一种新的工作方式。

Smarter manufacturing —— *Top-class office integration*

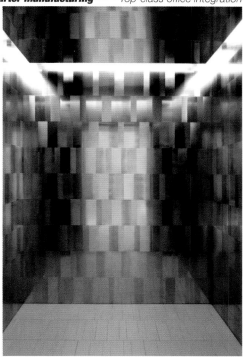

At the south side of the "New Sanlitun" project designed by us, a city development project "Sanlitun SOHO" is planned with a building area of 460,000m2. For its office area, we are responsible for the internal design. First of all, we design the sample room based on the overall design guideline.

In the sample room, some slender stainless-steel cupboards are hung irregularly on the wall, which highlights the room's deep shape and makes the space felt dynamic. In addition, the cupboard inside is pasted with coriaceous materials in four colors, showing fashion and gorgeousness. The application of this kind of color aims to fit the overall atmosphere at Street Sanlitun.

The partition is made up of transparent acrylic rods which is able to distort the projection, separating the office area softly and gracefully. The working table takes a curve shape, matching the line of outer wall.

We advocate a new working way in the non-territorial office system, at the street block leading the trend.

Smarter manufacturing — *Top-class office integration*

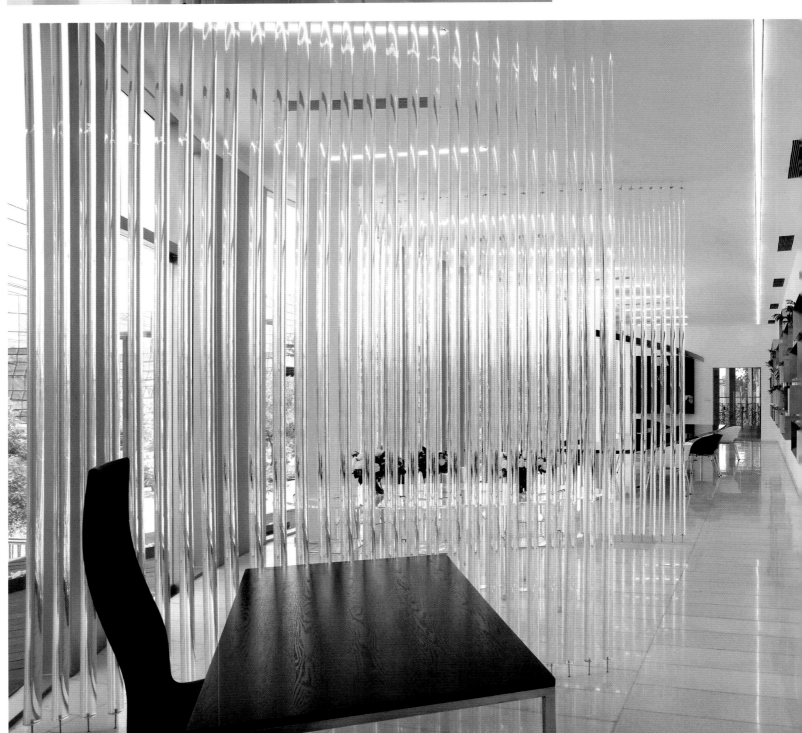

Smarter manufacturing —— *Top-class office integration*

佳敏企业办公室
Office of Carbing Enterprise

设计单位：隐巷设计顾问有限公司
设　计　师：黄士华 MAC HUANG、孟羿彣 CARRIE MENG、袁筱媛 EVA YUAN
参与设计：藏弄设计团队
项目地点：台北南港
建筑面积：约 660 m²
主要材料：德国拜耳透明中空板、指接柚木实木、木丝水泥板、3mm 木色夹板、黑色烤漆玻璃、白色烤漆板、PVC 地板、12mm 透明钢化玻璃
摄 影 师：王基守

　　基地座落于台北南港闹区，位于大楼B1空间，采光稍差。

　　第一次勘查场地时，感受到的是机械、黑油气味混杂的空旷，不良的采光，即使是大空间，却有50%消失在黑暗中。旧有的管道错综复杂的交错在天花上，整理归纳是我们的首要任务，赋予空间新的灵魂。

　　业主希望新的办公室能为企业带来新的精神与品牌定位，办公室功能区分为几大类，接待区、会议区（发表说明）、办公区、展示区、教育训练区（培训教室）以及仓储区。考虑办公与展示重叠使用的空间，产品展示的开放性与商业机密的私密维护相互冲突，矛盾的交叉点即是我们的设计理念。

矛盾的交叉点即是设计理念——Crossed the cube

　　入口接待区2000mm高之弧形墙面刻意与天花板保持区隔，利用高度落差让大会议室与接待区光线互通，同时保留会议室之私密性，墙体颜色采用企业标准色涂装，强化企业识别，大地色彩也为空间增添活泼感，地面PVC地面45°铺贴是入口设计的延伸，我们希望空间有更高的自由度与延伸，大会议室是

Smarter manufacturing —— Top-class office integration

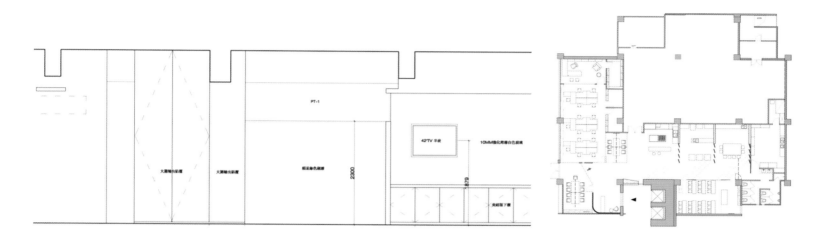

第一个立方体,会议室与办公区域特意留出三角形的区域,玻璃隔间在此处产生转折,并利用天花板延伸至地面设计,形成逆时针的U型,第一个立方体焉然成型。会议室内墙面材料为3mm木夹板,原有旧墙因整修不易,故将新做结构外露,一方面可以节省成本,一方面让不起眼的材料透过恰当的比例与处理,成为主轴,凸显材料质感。而办公区域则在两个会议室之间,将办公人员与来访客户隔开。

第二个立方体,是穿越小会议室至展示区的空间,包括通往仓库的门,方正的小会议室3面为透明清玻璃,利用结构梁,以上为板状隔间,以下为透明清玻璃,让不同材料立方体产生连结,主墙面为指接柚木实木,地面PVC铺贴方式自成一格。

展示区依据产品分成3大区,地面材料颜色界定出产品展示空间,倒L型中空板是为解决不同产品属性却需要保留空间通透性、加上照明布局的考虑所产生的解决方案,中空板有重量轻,透光性佳,钢性高等特点,所以我们放弃最初使用玻璃的想法;前面提到需考虑通透性却不彼此干扰,我们在平面图上计划135°的中空板隔间设计,从地面至天花的延展,跨越公共走道,依据地面PVC的铺贴角度,让彼此有同样的逻辑,而空间至此开始产生透视变形,即为第三个立方体与第四个立方体。中空板隔间光线从地面往天花投射,主要是利用板材产生的眩光产生隔屏效果,同时提高光线在空间中的延伸。展示区背墙面材料为木丝水泥板,灰色属于中性,色彩重量轻,不会影响产品陈设,同样的,我们也考虑了成本,故将之当成完成面使用,至于选择木丝水泥板是因为展示区会有较多的水气。

第五个立方体,教育训练区,同样因为墙面整修不易,故延续了大会议室的材料,温润的木色让空间产生变化,不是呆板的水泥墙面,地面颜色区分了走道与教室,期望培训学员因环境而激发创意,使用活动玻璃隔间可以让空间利用更有弹性,当人数过多时可以在走道上摆座位而不影响教学等功能。

材料的拼贴分割、延伸,空间的重迭与交叉,以及原有天花结构与管线的整理让 Crossed the cube 的概念得以实现,为了凸显立方体的概念,将原始天花板漆上80%的黑,我们希望高度自由化与的空间,搭配相互交错的线与面,让企业产生创新精神,并提升企业质感,超越自我、再创辉煌。

Located at the Building B1 in the downtown of Nangang, Taibei, the base suffers from poor day lighting. Upon the first survey on the site, what is felt is openness with mechanical and black oil odor; due to the poor day lighting, 50% of the large space is immersed in darkness. The old pipelines cross in a complicated way on the ceiling, and therefore, our primary task is to make an arrangement to give the space a new soul.

The owner hopes the new office can bring the enterprise new spirit and brand positioning and the office functions are divided into reception area, conference area (for statement), office area, exhibition area, education & training area (training room) and storage area. In consideration with the overlapping use of space for working and exhibition, we lay our design concept on the junction of contradiction between openness in product exhibition and keeping of business secret.

The junction of contradiction is design concept – Crossed the cube

The cambered wall of reception area at the entrance is painstakingly designed to keep a distance with the ceiling, thus connecting the light of conference room and reception area through the height drop and keeping secrets of the conference room.

The wall is painted in the enterprise's standard color to strengthen enterprise identity. Floor color adds liveliness to the space; the PVC floor paved at an angle of 45 degrees extends the entrance design, which is exactly our hope to make the space more free and extended. The conference room is 1st cube. A triangular area is especially set aside in the conference room and working area; here the glass compartment makes a turning and the ceiling is extended to the floor, thus forming a anticlockwise U-shape; then the 1st cube is shaped out. In the conference room, the wall surface is made of 3mm thick plywood. It is not easy to repair the original wall, and therefore, the new structure is to be exposed; in doing so, on one hand, the cost can be saved, and on the other hand, the unremarkable material can become principal through appropriate proportion and treatment, thus showing the tactile impression. And the working area lies between two conference rooms, separating the employees from visiting clients.

The 2nd cube is an area extended to the exhibition area after the small conference room is crossed, including the door to the storage. Three sides of the small square conference room are transparent glass; with the help of structural beams, the upper part is plate-shape compartment and the lower part is transparent glass, thus connecting cubes of different materials. The main wall surface is of finger-joint teakwood and the PVC floor is paved in one figure.

According to products, the exhibition area is divided into three sections which are defined by use of the color of floor material. In consideration with different product properties, transparent space and lighting layout, inverted L-shape hollow plate is applied; such plate features light weight, good light transmission and high rigidity, and that is why we give up the initial idea of applying glass. As for the transparent space without disturbance to each other, we plan a 135-degree hollow-plate compartment on the plane graph; the extension from the floor to ceiling crosses the public corridor and allows same logic for each other according to pavement angle of PVC floor, thus the space produces an effect of perspective distortion; then the 3rd cube and the 4th cube are born. The light in the hollow-plate compartment is projected from floor to ceiling when partition effect is produced owing to the glare reflected by the plate and meanwhile the light is more extended in the space. The back wall in the exhibition area is of wood cement board; the board is in grey color and

such color features light color weight and will not affect product exhibition. Similarly, in consideration with the cost, we apply the wall as finished one; and the wood cement board is selected for there will be much water vapor in the exhibition area.

The 5th cube – education & training area applies the same material as that in the large conference room for the same reason – uneasy repairing of wall. The gentle wood color makes the space change; the concrete wall surface is not mechanized any more and floor color distinguishes the corridor from the classroom, expecting trainees are inspired under such environment. The movable glass compartment can make the space application more elastic, which means that in case of too many people, seats can be arranged in the corridor, thus ensuring the teaching and other functions will not be affected.

The combination, separation and extension of materials, overlapping and crossing of spaces, and the arrangement of original ceiling structure and pipelines, all these things together realize the concept of Crossed the cube. To highlight this concept, 80% of the original ceiling is painted in black color. In doing so, we hope that the highly-liberalized space together with the crossed lines and faces can help the enterprise produce initiative spirit, promote quality, exceed self and again create glory.

隐巷设计顾问有限公司办公室
Office of XYI Design Co., Ltd.

设计单位：隐巷设计顾问有限公司
设 计 师：黄士华 MAC HUANG、袁筱媛 EVA YUAN、孟羿彣 CARRIE MENG
项目地点：台北市大安区
建筑面积：100 m²
主要材料：水泥板模墙、外墙质感涂料、9mm 黑铁板、2.5mm 拉丝不锈钢、3mm 木色夹板、10 mm 桧木板、染色桧木板、大甘木木皮、白色密度烤漆板、PVC、12mm 强化玻璃、8mm 黑色强化玻璃、5mm 灰镜、水性白色烤漆、人造石、银狐理石、复古面印度黑理石、金属砖、岩砖
摄 影 师：王基守

台北市的巷弄里其实卧虎藏龙，有许多不起眼甚至可能隐密到你不知如何上门的商家，而我们的想法是希望表达如"隐巷"字面般，一种低调、实务、质朴的理念，一种柳暗花明又一村的创新。

基地是位元元于台北市大安区巷弄里的旧民宅，上世纪60、70年代纵长向的建筑，前为小于3m的巷弄，后为1.5m的防火巷，采光差；但我们喜欢那样的年代，喜欢时光停留的错觉。寸土寸金的台北市区，大部分人总是尽可能的放大住宅功能空间，我们拆掉旧有车库后，留出前院空间，置入景观植栽与生态，让转折的巷弄与密集的住宅呼吸，而公司同事上下班时能过渡转换心情，门前的塑型鸡蛋花、会议室外的白水木与生态池让夏天的台北绿意盎然，让冬天的台北散发禅意，搭配门上桧木的淡香，使材料本身就是一种设计语汇；拆除旧有雨遮，仅保留会议室的部分，新的外露H钢做为结构梁，支撑会议室屋顶，我们去工厂采购了剩余的废弃桧木料，裁切成50mm宽的木片，以45°的方式拼贴，这是公司LOGO的夹角角度，也向80年代流行的拼贴设计致上敬意，使设计回归到手工与材料质感，并非仅是追求创新。

室外墙面材料使用板模灌浆，主要是为了做防水处理，并赋予设计语汇，右斜的墙线与左斜的会议室玻璃，如果你站在路口隐约能看见X的交叉点，入口地面使用复古面黑色理石用来定义出室内外空间的界线，建筑原有结构柱采用黑色铁板包覆，使其慢慢氧化，契合那旧建筑的年代；会议室外墙采用强化玻璃，解决室内采光不足的问题，也是节能方式。

内部我们保留当初拆除打凿的痕迹，这是建筑生命的周期呈现，与新的材料产生冲突感，却同时互为搭配，依据功能区分共分为六区，会议室与员工训练区、MINI BAR需互为重叠，电视崁入材料

柜门中，使其能90°旋转，根据需求使用，会议桌由大甘木、玻璃与黑铁结合而成，搭配意大利品牌单椅，此区同时有塑料、镜面不锈钢、玻璃、木夹板、黑铁板与水泥相互冲突产生的协调空间，而书柜从平面也采用斜度处理，Mini Bar内置入热水机与过滤器，使用KOHLER厨房龙头，三节式的设计让机能更臻完整，上方镜面壁柜延伸至室外，在玻璃隔间区域转换为黑铁板，一是为增加室内外的连结，二是将隐藏墙电箱，黑铁与板模墙面呈现出建筑未加工特征。

中间区域是设计师讨论的基地，桌面使用剩下的桧木板，仅使用木油处理，保留许多结眼与虫蛀孔，此区的黑铁椅是我们找了许多地方，有着工业革命时期的机械与粗糙感，设计部与讨论区互为重叠使用，设计部桌面为两块悬空的12mm强化透明玻璃，使用X造型支撑铁件，上方吊柜围内务使用柜，柜门上的冲孔灵感来自于电影变形金刚，是公司专属的代号，由同事与公司的字母组成，右侧的黑色玻璃书柜是方便行政人员能直接看到会议区与室外。

Toilet内有淋浴间，使用透明玻璃搭配单向镜做为隔间，从内部能清楚看见外面空间，而同时能保留隐私，墙面上的灰镜崁入10吋屏幕，可以播放影片与作品，不锈钢台面崁入订制的人造石水盆，高度仅100mm，内部为斜板，让水流缓缓滑过具肌理的石纹面。

后方是主管区同时也是创意生产中心，也是接待客户的区域，后方天花采用透明玻璃，增加采光空间，后方靠落地窗户的柜体内设有掀床，让需加班的同事有地方能休息，悬空的黑色玻璃隔间与折门是为了保持空气流通，利用T字型黑铁搭配吊筋做为结构。

整体设计我们希望能呈现材料本身的质感，无论是平整的墙面与粗糙的铁板，或是手工感的木板与镜面玻璃，透过比例将原本产生许多冲突的空间与材料转化成空间里的主角，低调却是空间的生命。

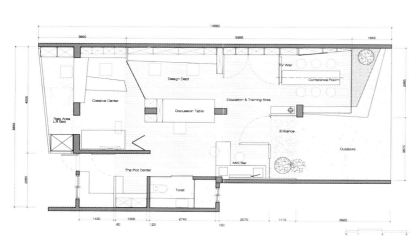

There are many capable companies which are often undiscovered in alleys of Taibei. They are even so secret that you couldn't find a way to them. We hope to express a low-key, practical and simple concept like the literal meaning of "hidden valley", that is, an innovation reappears after frustration.

The base refers to an old house in an alley of Da'an District, Taibei. It is a lengthwise building established in 1960s – 1970s. In front of it, there lies a valley whose width is less than 3m; behind it, there lies a 1.5m wide firefighting alley with poor day lighting. However, we like that kind of times for it gives us illusion of time staying. In Taibei where the land values a lot, most people will always enlarge the functional space of the house as much as possible. However, we will dismantle the old garage to leave room for the front yard where ornamental plants and ecological environment are set to make the turning alleys and dense houses breathe and employees convert their moods when going or getting off work. Frangipani in front of the door, messerschmidia argentea and ecological pool outside the conference room make the Taibei green in summer and Buddha-like in winter; all these things together with the light fragrance of hinoki on the door make materials themselves a kind of design language. We remove the old rain awning and only keep the conference room; the new exposed H-shape steel serves as structural beam to support the roof of conference room. We go to factories and purchase the remaining abandoned hinoki materials which are then cut into 50mm wide pieces and collaged in an angle of 45 degrees – exactly the included angle of the Company's LOGO; in doing so, we show our respect for the collage design popular in 1980s and make the design return to handwork and material sense, which is not just pursuing of innovation.

The outdoor wall surface is grouted with templates, aiming at waterproofing and showing design language. The wall oblique rightwards and conference room glass oblique leftwards make you vaguely see the junction of X if you stand at the crossing. The entrance floor is made of retro black marble to delimitate the indoor and outdoor space. The original construction pillars in the building are covered with black iron sheets for gradual curing, thus corresponding to the times of the old building. The wall outside the conference room is made of tempered glass to make up for the insufficient indoor lighting, which is also a kind of energy conservation.

For the interior, we retain the removing and chiseling traces which show the building life cycle. These traces conflict with the new materials and meantime match them. The functional area is divided into six among which the conference room overlaps employee training area and MINI BAR. The television is embedded in the cabinet door for a 90-degree rotation and use according to demands. The conference table is made from large sweet wood, glass and black iron, matching the chair of Italian brand. Here is a coordinated space formed by the conflict among plastic, mirror finished stainless steel, glass, plywood, black iron sheet and cement. The bookshelf is made with an oblique face. Mini bar is equipped with water heater and filter, and besides, KOHLER kitchen tap is applied, whose three-section design makes its function more complete. The upper mirror cupboard is extended to the outdoor and converted to black iron sheet at the glass compartment to make the indoor and outdoor in a better connection and hide the electrical wall box; the black iron and template wall surface present the rough texture of the building.

The central area is a base for designers to discuss. After application for table, the remaining hinoki boards are only treated with wood oil for retaining of eyes and holes. The black-iron chairs, which we find after walk-

ing around many places, are accompanied with mechanical and rough texture in the industrial revolution period; they are in overlapped use in the design department and discussion area. The table in design department is made of two dangling 12mm thick transparent tempered glass, with the iron part supported in X shape. The upper hanging case is applied for internal affairs; the punching holes on the case door, with the inspiration coming from the film Transformers, represent the Code exclusive to the Company, which are composed of letters for colleagues and the Company. The bookcase of black glass at the right side is set to make administrative staff directly see the conference area and the outdoor.

The toilet is equipped with shower; the compartment is set with transparent glass and one-way mirror, and in so doing, the outside can be seen clearly from the inside and meantime the privacy can be well kept. A 10-inch screen is embedded in the grey mirror on the wall to play films and works. The customized man-made-stone basin embedded in the stainless-steel table is 100mm high and equipped with chute board inside to allow water to flow over marbled surface with texture sense.

The rear refers to management area and creation center as well as reception area. The ceiling here is made of transparent glass to increase day-lighting space. The cabinet inside near floor window is equipped with a liftable bed for colleagues working overtime to rest. The dangling folding door and compartment of black glass, which are structured by T-shape black iron and hanging bar, aim to keep good air circulation.

The overall design, we hope, can present the material texture. Smooth wall surface and rough iron sheet, or craft-style board and mirror glass, all these things, through proportion, help the space and materials full of conflicts convert into leading roles in the space, showing low-key space life.

Groupm 群邑北京办公室
Office of Groupm Beijing

设计单位：**Mi2北京**
设 计 师：陈宪淳
项目地点：北京
建筑面积：**1200 m²**
主要材料：灰镜、单反玻璃、钢化玻璃、水切割铝板、穿孔喷绘、宇宙灰石材
设计时间：**2011**
摄 影 师：孙翔宇

Smarter manufacturing —— *Top-class office integration*

Groupm 从属 WPP 集团,下辖 Mediacom、Mindshare、MAXUS、MEC 等国际知名的 4A 广告传媒公司,这是北京的办公室,简单流畅的现代开放办公风格,巧妙利用全新的工艺赋予灰镜、铝板、石膏板、彩色喷绘等传统材料全新的生命力,空间结构实用至上,好用然后好看。这就是商业办公空间最永恒的出发点。

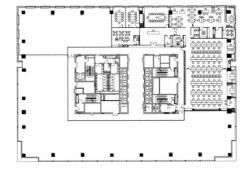

Groupm is subjected to WPP Group, affiliated to Mediacom, Mindshare, MAXUS, MEC and other international well-known 4A advertising media company. This is Beijing Office. Simple and smooth modern open style office, the designer has used the new technology process on the gray mirror, aluminum, Gypsum board, color injet and other traditional materials with new vitality. The spatial structure puts practical first and then focuses on the good-looking. This is the enternal starting point of the commercial office space.

Smarter manufacturing —— Top-class office integration

Smarter manufacturing —— Top-class office integration

道和设计机构办公室
Office of Daohe Design Institution

设计单位：道和设计机构
设　计　师：高雄
参与设计：吴运棕
项目地点：福建省福州市台江区世贸天城俪园二号楼一单元1606室
建筑面积：90 m²
主要材料：白色烤漆玻璃、乳化玻璃、镜面不锈钢、银镜、蒙古黑火烧面、壁纸、皮革、地毯
摄　影　师：施凯、李玲玉

　　办公室历时15个工作日的设计施工，坐落于福州市台江区儿童公园路世贸骊园，算是市中闹区，雏形初成。偶有好友造访，疑问不断。

　　你一直问，这儿为何到处镜子和玻璃，让你找不到门也找不到北。我笑而不答。浮华之上，孤独是造化所生，太阳所赐，只要行走，便如影般随形。而镜子一反常态，让光影找到跳跃的舞台，即使匆匆而过，也曾留下印记。这样，孤独便安静了，随我一同居隐于市。

Located at Shimao Liyuan District, Children's Park Rd., Taijiang District, Fuzhou, which can be considered as downtown of the city, the office is constructed for fifteen working days and then takes an embryonic form. Sometimes, friends make a visit with questions.

You always ask, why mirrors and glass are everywhere here, making it hard for you to find the door and the direction. I smile and make no reply. Above vanity, the loneliness comes from predestination; what the sun grants will follow you as a shadow as long as you walk. While the mirror overturns the normality to allow the shadow to dance on its own stage; even if it passes through in a hurry, its imprints have remained. Thus, loneliness becomes quiet and hides in the city together with me.

东灿五金贸易有限公司全球总部
Headquarter of Dongcan Hardware Trading Limited Company

设计单位：十方圆国际设计工程
设 计 师：赖建安
参与设计：高天金、陈书芳
项目地点：中国上海市松江区小昆山广富林路4855弄99/100号
建筑面积：3850 m²
主要材料：老洋松集成材、钢构、大理石、玻璃
设计时间：2011.01~2011.03
施工时间：2011.05~2011.09
摄 影 师：深蓝摄影工作室
撰　　文：陈书芳

一层平面布置图

本案引入"cabinet construction"箱体建筑理论，让室内空间的相互渗透，在视觉感觉形成"借景"、"对视"的效果，造成"空间停泊点"。悬浮在空中的红色箱体，为这种效果创造了有利的机会。使室内空间融入原建筑中，成为相互的视觉冲击点，让彼此的出现成为彼此记忆的出发点，空中交错玻璃廊桥是设置这种"停泊点"的重要手段。这种意境被现代著名诗人卞之琳的《断章》充分的诠释：你站在桥上看风景/看风景的人在楼上看你/明月装饰了你的窗/你装饰了别人的梦。

细节设计是对建筑形态和空间认知方式的深化，透过材质加以诠释。本案运用铁件、石材、玻璃、镂空木隔栅等，将材料最本质的肌理经创意提炼，并呈现出来。让人触及该空间时能体验到材质经冶炼后，体现原生态的价值感。

归零的思考是本案设计师在此次设计的最大尝试和突破。寻求在摒弃的过程中，淬炼本源才是设计的初衷。

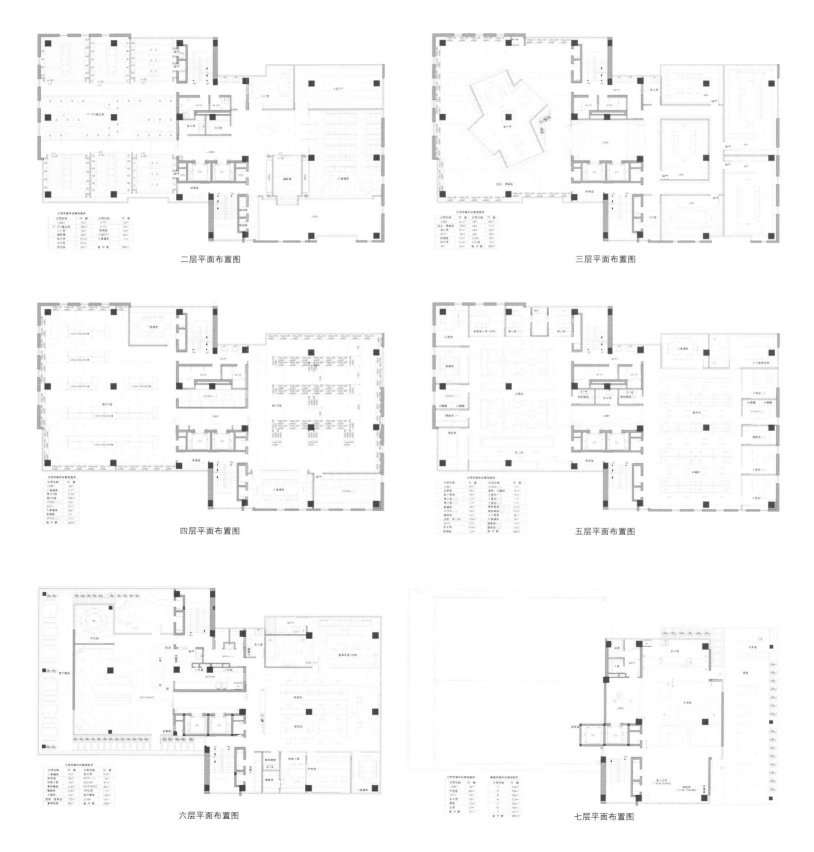

二层平面布置图

三层平面布置图

四层平面布置图

五层平面布置图

六层平面布置图

七层平面布置图

This case has introduced "cabinet construction" construction theory, letting the interal space interfaced with each other. It is mainly to form "introducing scene", "facing with each other" effect in terms of visual, forming a hanging mooring point effect. The red box suspended in the air, has created a favorable opportunity for this effect. The internal space is merged into the formal construction, which has becoming the impact visual point, letting the appearance of each other become their memory starting point. The intricate glass corridor is an important means of such "parking point". This mood is interpreted fully by modern famous poet Bian Zhilin's "Duanzhang", that is, you stand on the bridge to enjoy the scenery/the people who is enjoy scenery is looking at you at the upstairs/moon has decorated your windows/you have decorated other people's dream.

The details design has deepened the architectural form and spatial cognition through materials. This case adopts iron, stone, glass, hollow wooden grille, etc. Refining the most essential texture of materials and showing to people, let people could experience the original ecology value when entering this space.

Back-to Zero is the biggest attempt and breakout of this case. Pursuing among the abandon process, refin the original spirti is the purpose of this case.

Smarter manufacturing —— Top-class office integration

瑞士苏黎世谷歌EMEA工程中心
Google's New EMEA Engineering Hub in Zurich, Switzerland

设计单位：CLIEN
设 计 师：Stefan Camenzind、Tanya Ruegg-Basheva
项目地点：苏黎世
建筑面积：12,000m²

新建苏黎世谷歌EMEA工程中心项目需要考虑个人办公空间的功能性与舒适性及公共区域的选择性与多样性，从而为苏黎世谷歌员工创造一个健康舒适的工作环境。承接此项目的建筑师将研究的重点放在与苏黎世谷歌员工密切相关工作环境的各个方面。已经竣工的谷歌大楼真实地反映了公司是一个具有前瞻性理念的企业，愿意采纳建筑团队提出的创新方案，以促进苏黎世员工的才能和创造性。

一层平面布置图　　二层平面布置图

Smarter manufacturing — Top-class office integration

The new Google EMEA Engineering Hub in Zurich is about functionality and flexibility in the personal workspace and choice and diversity in the community areas, creating an environment that holistically supports the Zooglers in their work and well-being. The research undertaken by the architects ensured their efforts were sharply focused on aspects of the environment that mattered most to the Zooglers in their working lives. The completed Google building is a true reflection of a forward-thinking company that was prepared to adopt innovative approaches proposed by the architectural team to determine the optimal design concept to nurture the talent and creativity of their Zurich employees.

Smarter manufacturing —— Top-class office integration

123

汉诺森设计机构办公室
Office of Hallucinate Design

设计单位：汉诺森商业空间设计事务所
设 计 师：王文亮
项目地点：中国深圳
建筑面积：**500 m²**
主要材料：环氧树脂自流平、清水混凝土浇筑
摄 影 师：汉诺森设计机构

在500 m²空旷的空间内，汉诺森以3条平地而起的泥管道的设计，分割出各设计部门所需空间，并串联起各个空间：包括前台、洽谈区、摄影棚、会议室、客服部、生产部、空间设计部、平面设计部、行政财务部、总经理办公室、洗手间、仓储处。在室内设计的导向上，提供了区域引导的功能并带来空间游走的乐趣。

除了保留原有天花，内部空间的设计我们全部采用清水混凝土浇注作为主要材质，唯一的化学材料是自流平水泥做基础的环氧树脂地面，映衬出明净、整洁的工作氛围。

身为设计师，总是免不了长久地置身于公司以孕育灵感，我们提供最舒适和惬意的办公环境：henman miller的人体工学椅、flos温暖简洁的落地灯、kartell & stark的Ero椅等，在这样的空间设计里，让每个同事都切身感受到优秀设计的价值。

In the open space with an area of 500m², Hallucinate partitions, with three mud pipes rising from the ground level, the space to meet the need of each design department and then connect the spaces including reception area, conversation area, studio, meeting room, customer service department, production department, space design department, planar design department, administrative and financial department, general manager office, washroom and store room. The office is designed with a function which is known as region guiding and able to bring fun during walking in the space.

As for the design of space inside, the original ceiling is remained, and concrete with water will dominate in the space with the only chemical material - epoxy resin ground which is based on the self-leveling cement and reflects out the clean and tidy working atmosphere.

It is unavoidable for a designer to long stay in the company to seek for inspiration. Therefore, we provide the most comfortable and pleasing working environment: ergonomic chair of Henman Miller, warm and simple floor lamp of Flos, Ero chair of Kartell & Stark, etc. In such space, every colleague will feel in person the value of this excellent design.

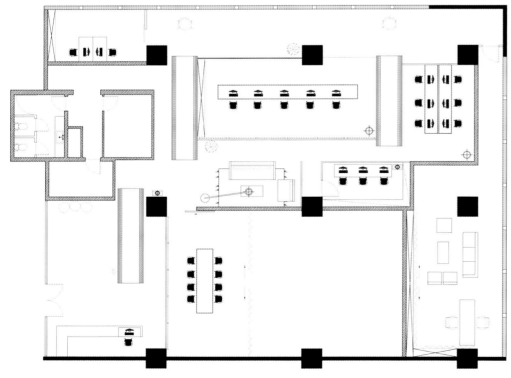

Smarter manufacturing —— *Top-class office integration*

红坊办公室
Red Town Office

设计单位：塔然塔建筑设计咨询(上海)有限公司
设 计 师：**Enrico Taranta、Juriaan Calis、Giorgio Radojkovic**
项目地点：中国上海红坊雕塑园
建筑面积：**120 m²**
项目时间：**2010**
摄 影 师：申强

塔然塔建筑设计咨询(上海)有限公司在中国上海设计了自己的工作室。设计反映了工作室内不曾间断的创意进程,旨在提供可支持一系列日常功能的适应性空间。

办公室原为一间金属工厂。工作室原计划建成两层,但切割的对角线钢结构给工作造成了麻烦。结构与天花距离太近,无法对办公室内地板和桌子进行传统布局。因此,在钢结构的正上方铺设地板。在钢材剖面之间的空间设立4个工作站。这样,地板就转换成了连续的办公桌,而4个嵌入式的工作站可用作传统的办公间。利用大型的"工作台面"创造出的开放空间,设计师可尽情思考、绘图、讨论、起草、制模、就座和放松。办公空间的这种布局便于不同项目和学科之间在工作室内进行交流。

低层的各个工作站沿着窗户放置。绿色的雕刻桌可用于公共活动。不拘一格的波形中央楼梯让人产生错觉,觉得一大滴水正要从天花落下。一走上楼梯就淹没在高度饱和的亮红色中,使得办公区域对比更加鲜明。此色调在二楼再次使用,以突显嵌入式的办公区。

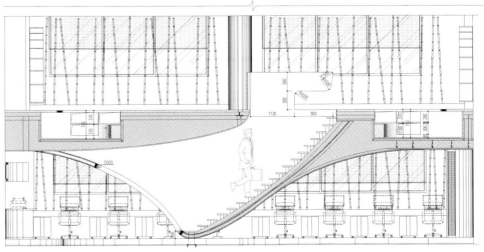

Smarter manufacturing —— *Top-class office integration*

Shanghai based practice Taranta Creations has designed their own studio located in Shanghai, China. The design is a reflection of the ongoing creative process within the studio. The intervention seeks to provide an adaptable space that supports a range of informal functions.

The office is situated in a former metal factory. The excising diagonal steel structure was causing difficulties to fulfill the wish of the studio to create two floors. The distance between the structure and ceiling was too less for a traditional office layout of floor and tables. Therefore a floor is created just above the steel structure. Four working stations are placed in the space between the steel profiles. This way the floor is transformed to one continuous desk, while the four recessed stations provide a more traditional workspace. The large 'work floor' invites the designers to use the open space for thinking, sketching, meeting, drafting, modeling, sitting and relaxing. This informal interpretation of office space encourages cross-pollination between the different projects and disciplines occurring within the studio.

On the lower floor the individual workstations are placed along the window. A green sculptural table can be used for communal activities. Informal and contoured, the central staircase is reminiscent of a large droplet of water ready to fall from the ceiling. Upon entering the stairway, a highly saturated environment of bright red engulfs and surrounds the individual, starkly marking the transition between the contrasting office areas. The color is repeated on the second floor as an accent to highlight the recessed work areas.

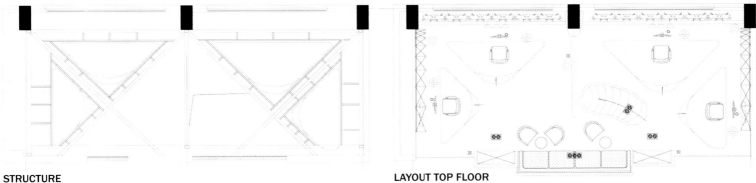

STRUCTURE **LAYOUT TOP FLOOR**

Smarter manufacturing —— Top-class office integration

红桃网办公室
Office of Aceona.com

设计单位：北京艾迪尔建筑装饰工程有限公司
设 计 师：黄丽元
项目地点：北京中环世贸中心D座7层
建筑面积：**300 m²**
主要材料：白色微晶石、红色夹胶玻璃、木地板、地毯、乳胶漆
设计时间：2011

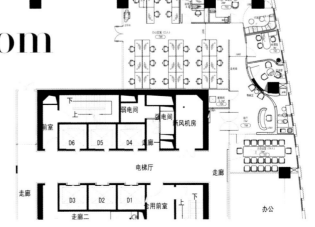

红桃网的这间办公室面积虽只有300 m²,但作为中国互联网行业首个"推荐式高端人脉资源整合平台",空间的设计体现了互联网企业的特性和高端资源互动的精神,营造出一种极具特色创意又不失舒适安宁的办公交流空间。

窗体底端、前厅、等候区的墙面设计成一条有趣的曲线,将VIP体验区与入口贯穿在一起,形成红白相间的舞动长廊。这种连续的流线型墙壁是将接待台、LOGO板、洽谈吧台、休息区洞口、经理室墙面等组织在一起的。墙壁上多重开口又与天花及地面造型相呼应,形成有趣生动的弧面及层次。白红两色对比鲜明,自由线体组合的造型墙更衬托出主体曲型墙面的动感和飘逸。

作为接待高端VIP客户的洽谈室,柱面墙面采用了背漆玻璃与镜子这种具有反光特性的材质,使得狭小的空间得以延展,拐角处突出的柱子也得到很好的隐藏。

办公室采用了大面积白色作为主基调,给人们形成视觉上的扩张和延伸,同时点缀的红色让空间显得更加活跃。

As the first "recommended high-end resources integration platform" in China internet industry, Aceona.com owns an office which reflects both features of internet industry and spirit of high-end resource interaction in spite of the small office area of 300m² where a working and communicating space is built with unique creation, coziness and quietness. Window Bottom

The walls of front hall and waiting area are designed into an interesting curve, which connect VIP experiencing area and entrance to form a long corridor in red and white colors. This continuous streamlined wall is in connection with walls of reception desk, LOGO board, discussion bar counter, resting area entrance and manager room. The multiple openings on the wall echo shapes of ceiling and floor, forming an interesting and vivid cambered surface and level. In addition to the white and red colors with strong contrast, the wall with combination of free lines more highlights the dynamic and elegancy of the curve surface.

As for the conference room to receive high-end VIP customers, cylindrical wall surface applies painted glass and mirror with reflection property to make the narrow space extend and hide the conspicuous pillars at the corner.

The white color dominates in the office, bringing people a visual feeling of expansion and extension; and a little bit of red color makes the space more vigorous.

思维繁殖场
Idea Breeding Ground

设计单位：建构线设计有限公司
设 计 师：沈志忠
项目地点：台北市
主要材料：水泥粉光地坪、橡木实木皮、钢筋与铁件成型喷漆、石英石、W贴法铁刀实木海岛型木地板、砖造型墙
建筑面积：**215 m²**
设计时间：**2008.10~2009.01**
施工时间：**2009.01~2010.01**

Smarter manufacturing —— *Top-class office integration*

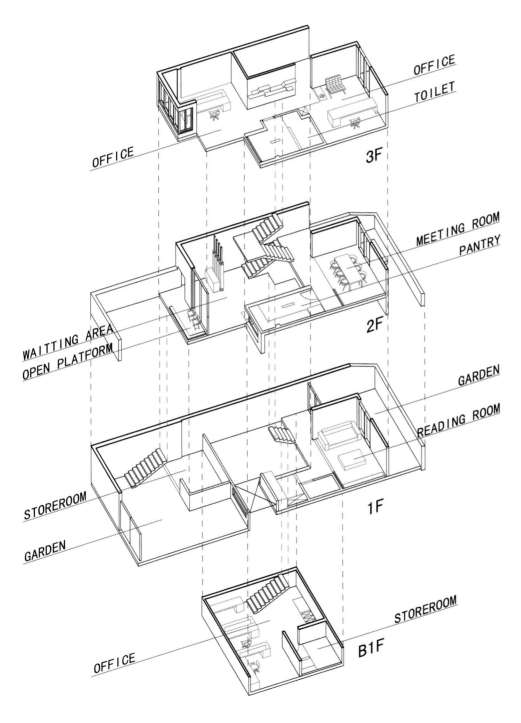

　　空间设计理念以思想繁殖场为设计主轴,以"生成繁衍"的概念创造不同以往的办公室型态。其目的在于让空间中的每一个成员在此思维场域中放松思考,并产生互动与对话,让想法在无拘束的空间中不受限地繁衍分裂与结合生长后,终而得以释放出独特的设计能量。

　　在建筑的前院以樱花及樟树围塑出具有宁静特质的场域。穿过一道楼梯进入室内后,玄关摆放4根枕木做为通往二楼楼梯与入口的隔屏,同时也成为大厅视觉焦点所在。挑高大厅由延伸至二楼的白墙、红砖与玻璃所围塑而成,并与前后院的景致结合,将视线的垂直与水平延伸感发挥到极致,而质地纯粹的清水模地板呼应红砖的朴实与木头的温润,强化空间的触觉深度。大厅旁的折梯做为室内的垂直动线,不同阶数的楼梯将工作室、会议室、书房等每个单元空间区隔开来,也将彼此串连起来。单元空间以活动拉门定义属性,依照不同的使用时机使室内产生不同的表情,不但为空间创造更多使用上的可能性,也让人激发出更多元的灵感。

B1层平面布置图

一层平面布置图

二层平面布置图

三层平面布置图

The design of the space is based on the philosophy of Idea Breeding Ground; hence, it results in the office style different from the past by the concept of "breeding". It aims to allow every member in the space to think and relax in this idea ground, and interact and have dialogue with each other. Thus, ideas can be generated, separated, and combined, and grow in the free space which in the end will release a unique designing energy.

The front courtyard of the building is enclosed by cherry blossoms and camphor trees to create a space with a tranquil characteristic. After passing through the stairs and entering the building, you can find four timbers placed in the hallway as the screen to separate the space between the stairs to the second floor and the entrance. They are also the visual focus of the hall. The high-ceilinged hall is enclosed by white walls, red bricks, and glass, of which, all are

extended to the second floor and combined with the landscape in the front and back courtyards. It is the extension of vision to the horizontal and vertical extremes. Fresh water formwork floor responds to the humbleness of red bricks and the mildness of wood, and reinforces the sense of touch. The folded stairs beside the hall direct the vertical indoor flow. Stairs with different numbers separate and connect the unit spaces such as studios, conference rooms, and study. Attributes of unit spaces are defined by sliding doors; according to the different timings of usage, there will be different indoor expressions. It not only creates more usage possibilities of the space, but also elicits more diverse inspiration.

Smarter manufacturing —— Top-class office integration

143

Smarter manufacturing —— Top-class office integration

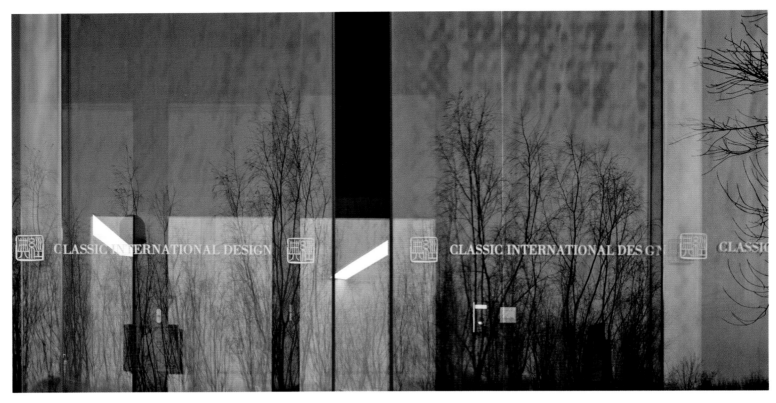

经典国际设计机构(亚洲)有限公司办公室
Office of Classic International Design Agency (Asia) Co., Ltd

设计单位：经典国际设计机构(亚洲)有限公司
设 计 师：王砚晨、李向宁
项目地点：北京市朝阳区奥林匹克森林公园南园内 2A-05
建筑面积：**800 m²**
主要材料：白色乳胶漆、透光膜、钢板、白色人造石、玻璃、白色铝塑板、自流平地面

"慢"是"快"的基础，只有习惯"慢生活"，才能够快速准确找到定位，而不会迷失自己。要慢下来(Slow)，是因为"快"让人错失了很多美好的事物。所以我们倡导：慢生活、慢餐饮、慢睡眠、慢工作、慢情爱、慢社交、慢读书、慢运动、慢音乐、慢设计……

我们秉承这一理念，将"慢"的理念延续到我们的办公空间。首先是选址，北京的空气净化器——森林公园成为我们的最佳选择。这栋建筑位于森林公园的腹地，依山傍水，独立清幽，完全隔绝都市的繁杂和喧闹，是真正意义上的室外桃源。

我们尽量尊重原有建筑的空间结构，错层、高达6 m的空间高度、三角形的采光顶，都得以保留，原有的不规则结构梁成为我们的照明基座。狭长的残疾

人坡道成为材料区和文印中心。而宽敞的户外露台成为绝佳的休闲和放松的区域。

我们使用最单纯的设计语言，把安静的气氛融入空间之中，置身室内，窗外的自然美景是最大的视觉重心。我们可以静观微风吹过，枝叶飘摇。水光天色，山重树茂，无不快哉。室内家具和艺术品的选择也同样遵循"慢设计"的理念，只有被称之为经典的才能称为空间的主人，明式圈椅、The Chair、Y Chair、Ghost Armchair轮番登场，Pop Art、北魏造像、当代艺术交相辉映，共同谱写一组和谐的乐章。

置身这样的空间之中，心会自然地安静下来，快的节奏和习惯会慢慢远去，我们会更清晰地思考，更深入地研究，以致更精准地处理设计中的所有关系，努力创造更具深度的作品。为中国设计走向世界贡献自己的微薄之力。

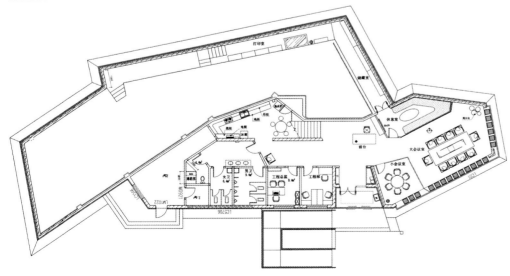

一层平面布置图

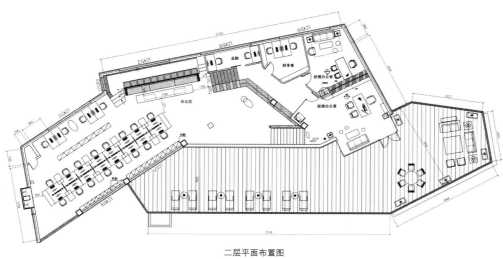

二层平面布置图

"Slow" is the basic of "fast". Only used to "slow life", can you be able to find your position quickly and accurately without losing yourself. We shall slow our steps, because "fast" let us miss a lot of beautiful things. Therefore, we call on slow life, slow food and beverage, slow sleep, slow work, slow love, slow communication, slow reading, slow sports, slow music and slow design.

Upholding this concept, we put "slow" into our office area. The first thing is to choose address. Beijing's air purifier, Forest Park is our best choice. This building is located at the center of the Forest Park, surrounding by mountains and waters, independent and quiet, which has isolated from the bustle and hustle city. It is the true outdoor place for people.
We try our best to respect the spatial structure of the original building. The overlapped, 6 meters high, triangular skylights are all preserved. The irregular structure beam has formed our light base. The long and narrow dis-

abilities road becomes the material and cultural printed center. However, the outdoor terrace has become the perfect leisure and relaxation areas.

We are using the most simple and purity design language to put the quiet into the space. Standing in the room, the nice landscape of outdoor is our biggest visual center. We can enjoy the gentle breeze that makes leaves swaying. Water and sky, mountain and trees, how joyful it is! The interior furniture and art products are all in line with the "slow design" concept. Only the classical can be called the owner of the space. The chair, Y Chair, Ghost Armchair are walking on the stage one by one. Pop Art, Beiwei Dynasty statues, contemporary art are interacted with each other, showing a harmonious piece of music.

Standing among such space, our heart will be calm down slowing. The fast speed life and habit will leave us gradually. We will have a clearer meditation and deeper research to handle all the relationship of the design to create a better works.

某平面设计公司办公室
Office of a Planar Design Company

设计单位：**SGM SPACE** 设计咨询机构
设　计　师：上官民
建筑面积：**75 m²**
摄　影　师：**SGM**

Smarter manufacturing —— Top-class office integration

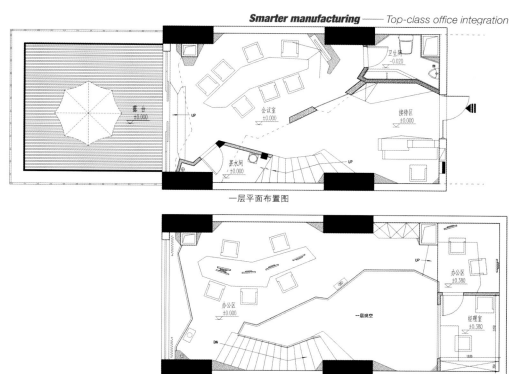

一层平面布置图

一层挑空平面布置图

　　本案是一个仅有45 m²，层高5 m的长方形盒子空间。作为一个平面设计公司的办公空间，业主希望看到一个时尚、独特、与众不同的办公场所。

　　经过设计整合，空间许多边界的直角都改变成锐角或钝角，形成一种不规则的边界形态。为了满足实用功能需求，在5 m的空间里搭建了一个夹层。而夹层和楼梯的边界也是不规则的，这样形成的挑空区域也是不规则的。如此的改变让一个普通的长方形盒子变得有趣、生动起来。这种手法甚至延续到会议桌，办公桌等家具设计上。形成一个统一的整体。

　　为了增强视觉冲击和强化公司形象，选用公司的VI形象色——橙红色。采用这种颜色的有机玻璃切割成不规则的块状装饰在空间的不同角落。它像"彩带"一样把上下空间紧紧的联系在一起，看似不规则，但又有内在规律。白色与橙红色的搭配看起来更纯粹和生动。一个时尚、独特、与众不同的全新办公室诞生了……

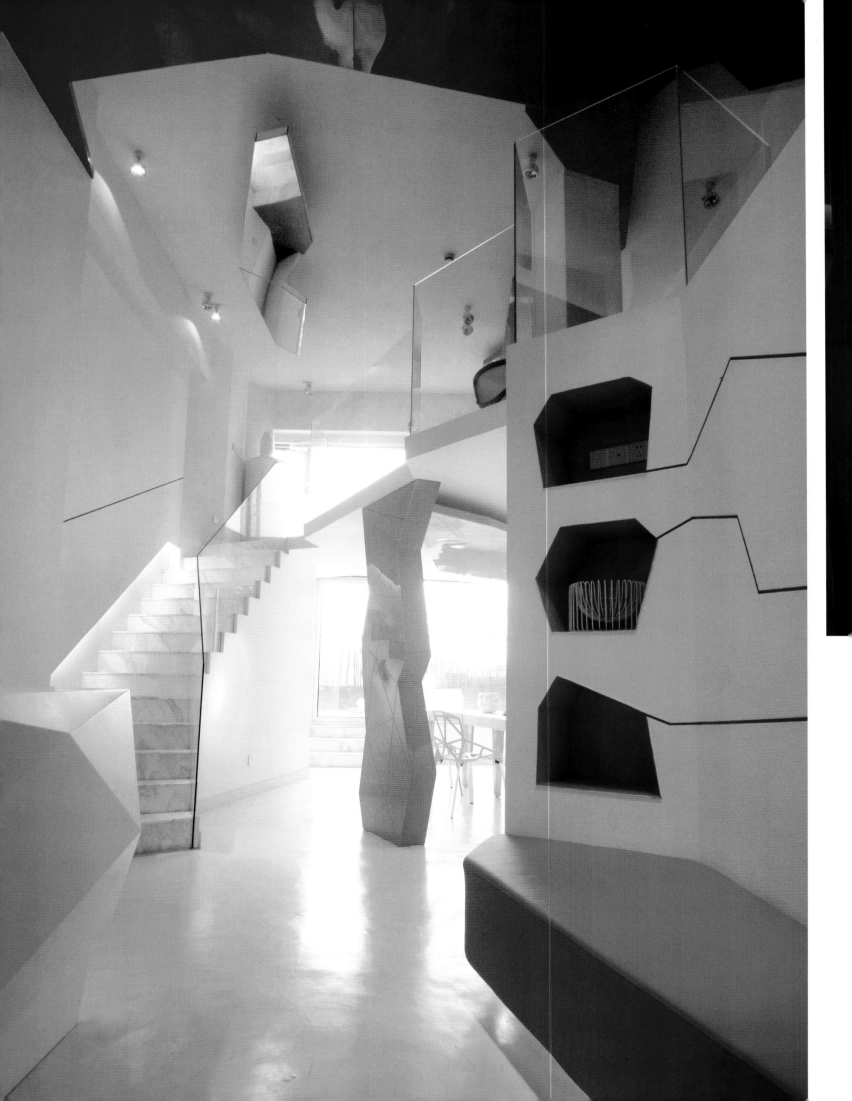

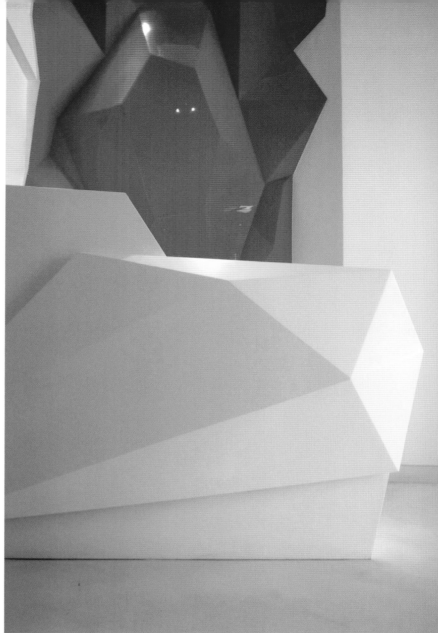

This case refers to a rectangular five-floor box space with an area of only 45m2. For the company is engaged in planar design, the owner hopes that the office space is fashionable, unique and out of the ordinary.

After space integration, the right angles of many boundaries in the space are changed into acute or obtuse ones, forming irregular boundaries. To meet the demand for practical functions, an interlayer is built within a 5m space; the boundary between interlayer and staircase is also irregular, resulting in an irregular area. Such changes make a common rectangular box become interesting and vivid. This kind of technique is even applied in design of conference table, desk and other furniture, thus realizing uniformity.

To enhance visual impact and strengthen company image, the VI image color – orange red is selected. The synthetic glass in such color is cut into irregular pieces to decorate different corners in the space. It is like a "color ribbon" to connect the upper and lower spaces closely; it seems irregular, but in fact, the inherent law exists. White color together with orange red looks more pure and vivid. Then a fashionable, unique and unusual office is born…

雷迪有限公司办公室
Office of Leidi Limited

设计单位：设计集人（www.designsystems.com.hk）
设 计 师：林伟明、梁芬华、杨励莹、张星、王永健、
　　　　　张芷茵、方欢欢、李婉恩
项目地点：上海天山路780号
建筑面积：974 m²

这个办公室项目楼高5层,是一家地下管线管理及检测公司位于上海的办公大楼。大楼拥有逾40年历史,属典型旧式现代主义建筑。由于当年采用的是讲求快捷及低成本的旧式建筑方法,即先预制好大部分水泥组件,再于现场装嵌,所以,整幢楼宇的所有墙身皆作负重之用,所有间隔皆不得改动,亦因此对这个项目的空间配置造成限制。经过力学计算之后,我们在负重要求最小的顶层拆除部分墙身,并加设钢架承托楼顶。所有楼层维持原来"中央一条通道,两旁布满小房间"这个井然有序的间隔。

一座富有历史的建筑物是记忆的载体。项目所在地原为牛类养殖场,后来成为农业原料国企之办公大楼,外墙与室内装修已沿用多年。其位处住宅区,毗邻上海特色的民房。为了使大楼与周边社区及环境互相协调,设计师选择保留大楼原有的窗户、楼梯扶手栏杆、墙身等这些饱历风霜的历史见证作为框架,以简单的白色油漆进行翻新,让大楼在隐然褪色的岁月痕迹当中,细诉昔日往事。

这个办公室项目的客户是一家地下管线管理及检测公司,专门以先进的高科技仪器及方法如微型机械人等为水、电、热、通讯等各项公共事业提供地下检测服务,以便城市基建项目进行工程勘探和制定施工方案。由于这家公司的服务宗旨正是"看见你所看不到的",所以,设计师在这次的办公室大楼里特别设计

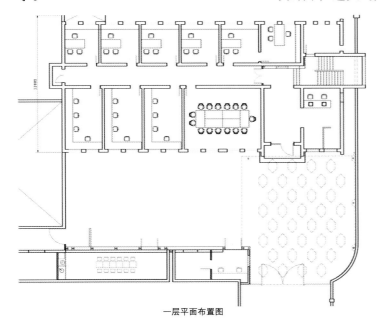

一层平面布置图

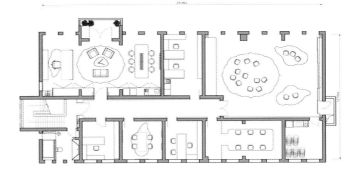

五层平面布置图

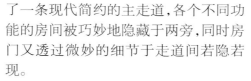

了一条现代简约的主走道,各个不同功能的房间被巧妙地隐藏于两旁,同时房门又透过微妙的细节于走道间若隐若现。

这个项目的设计意念是怀旧老上海,设计师以铜作为设计语言贯穿大楼各个楼层。所有标识指示、门框及装饰皆以铜特制,令空间在焕然一新的同时,淡淡地流露老上海的独特风韵。窗外景色尽是充满老上海情怀的旧式海派民居;怀旧的设计正好将室内空间与窗外世界联系起来。

由于大楼内的楼层高度和空间间隔较小,各楼层办公室的天花均以铝材特别制成不同造型,以符合功能上的声学要求。如会议室的凹凸造型冲孔铝板天花,既具吸音效果,又能避免使用者直接看到光源,从而令室内光线更加舒适。

客户是一家中国公司,但有不少外国的客户及供应商经常到访交流。为了强化这家公司的独特身份,董事长办公室内设置了一张云石茗茶桌,象征富贵的牡丹图案地毯,以及寓意平衡之道的铜制吊饰,来反映公司的文化背景与内涵。

This is a 5-storey office design project for an underground utility management and detection company in Shanghai.

This office block is a typical modernistic architecture of over 40 years old. It was built by assembling precast concrete slabs onsite, which was a popular building method at that time due to high efficiency and low cost. In this way, all the walls in the building are load-bearing, and no partitioning can be modified. This thus becomes the limitation on the space allocation of this project. After mechanical analysis, we removed some of the walls on the top floor and added a steel frame to support the roof. The tidy partition of having "one central corridor with small rooms on two sides" remains on all levels.

A building full of history is a vehicle of memory. The site was originally a cattle farm. It then became a national agricultural supplier's office building, of which the facade and interior has been used for years. The building is surrounded by typical Shanghai houses in a residential area. In order for the building to be in harmony with the local community and surroundings, the designer has chosen to keep historical elements like the original windows, staircase handles and walls as the framework

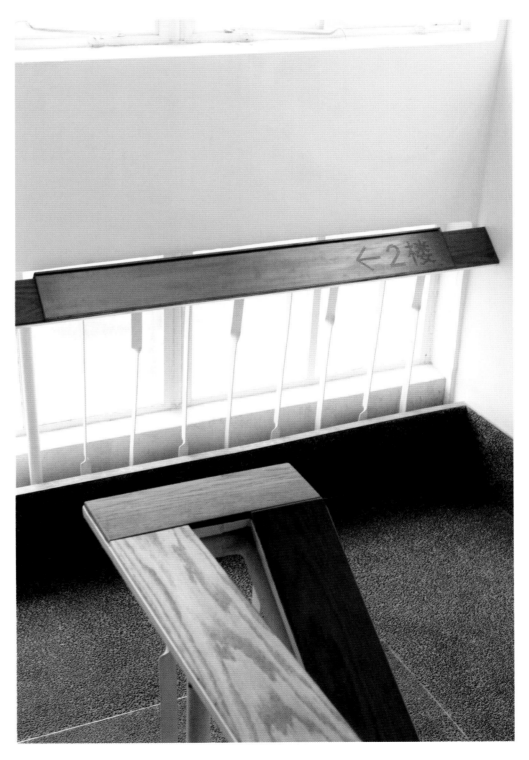

and repaint them white, so that the story of the building unfolds quietly through the pale traces of time.

The client is an underground utilities management and detection company. They specialize in providing underground detection services for public utilities and urban infrastructures by using high technologies like micro-robots. The company's philosophy is "seeing the invisible". Therefore in this project, a modern and minimal central corridor is specially designed in which the rooms of different functions on the two sides are neatly concealed, while the doors to these rooms are subtly revealed.

The idea of this project comes from Old Shanghai. The material copper is used as the thematic design language. All signage, door frames and decorations are made of copper to revitalize the space as well as to gently manifest the unique charm of Old Shanghai. Outside the window are typical old Shanghai-styled houses. The retro design exactly links up the worlds in- and outside the building.

Since the height of each floor and the space of each partition is relatively small, the ceilings of the office rooms are made of custom-designed aluminium profiles to functionally fulfil the acoustic requirement. For example, in the meeting room, the undulated ceiling made with perforated aluminium profiles not only absorbs sound waves, but also cleverly hides the light source from direct sight to produce comfortable illumination.

The client is a Chinese company with many foreign clienteles and suppliers visiting them. In order to strengthen such a unique identity of the company, details with Chinese characteristics are designed to reflect the company's cultural background and values, such as the marble tea table, the peony-shaped carpet that symbolizes "prosperity", and the copper hanging decoration that signifies "balance".

Smarter manufacturing —— Top-class office integration

167

Smarter manufacturing —— Top-class office integration

深圳派尚设计公司办公室
Office Space of Shenzhen Panshine Interior Design Co., Ltd.

设计单位：深圳市派尚环境艺术设计有限公司
设 计 师：李益中
项目地点：深圳市南山区侨香路香年广场
建筑面积：**430 m²**
主要材料：白色乳胶漆、灰色自流平、白色聚脂漆、白色人造石、白色烤漆玻璃、白色亚克力、铁板（喷白漆）

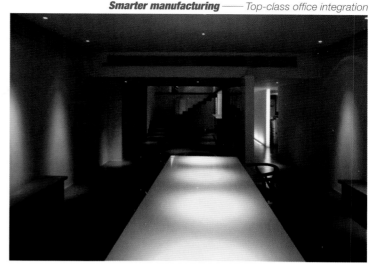

因为公司业务量不断增加，规模也不断扩大的关系，原有办公室无论在规模抑或是形象上都已经无法满足公司发展的需求，位于侨香路香年广场的新办公室由此而生。

原结构是一个有着4.8 m层高的挑高空间，除去梁高，净高4.2 m，如何在这样一个固有结构内设置夹层而不显压抑，是我们在设计中需要解决的问题。通过分析研究，我们选择纯净的白色作为空间主色调，配合合理的空间布局，精确的尺度考量，共同打造了一个极具结构感与线条感的洁净、大气空间。

色彩

白色作为整个空间的主基调，不仅界定了简约优雅的空间氛围，也增加了室内亮度，使空间变得开阔明亮，带给人愉悦的空间感受。灰色自流平地面，以及黑色门框，楼梯踏步的运用则打破了因过多白色显出的苍白平淡，活跃了空间气氛，同时也强化了空间的构成感。

结构

简洁流畅的线条，超薄楼板、钢板的设计，不仅从结构上勾勒出了空间简约优雅调性，也在一定程序上扩展了空间深度。楼梯踏步极具线条感的设计，成为空间造型的一部分。书架采用超薄钢板喷白漆设计，构造出简约的造型，同分隔图书区与设计区墙面漏窗相呼应。办公桌椅、电脑、灯具等的选择同样是对空间整体构造的一种延续。

空间布局

夹层的设置使空间得以合理规划。大堂以及办公区域上空的挑空布局，是整个空间特色之一，虽然减少了空间的使用面积，但通过挑空设计以及具有通透性玻璃材质的运用，扩大了空间内部景深，也连接了外部景观。同时对于员工来说，不仅具有更开阔的视觉空间，通过挑空区域互观楼上楼下动态，方便了相互之间的沟通交流。

一楼大堂与会议室之间90°可旋转墙的设置是空间布局中的另一个亮点。墙面与接待台垂直时，分隔出一个独立会议区域。会议时间以外，则可将旋转墙展开与接待台平行，大堂与会议室融为一体，透过会议室落地玻璃窗引入室外景观，增添了空间开阔感。

灯光

吊灯，壁灯作为空间主要照明设施，融合成空间造型的一部分。射灯的大量使用，则增强了空间的表现力。大堂接待台底面，背景墙地面以及侧面设置内嵌灯带，即完美解决了因层高低而造成的照明问题，又与整体设计语言相一致。

新办公室的设计不仅是对我们一直以来坚持有节制优雅设计理念的一种展现，也是对我们总结出的设计方法的一种实施。不追逐特定的设计风格，所有设计源于对现场结构、实际需求等的理性分析，创造出一个具有针对性的独特的设计作品。

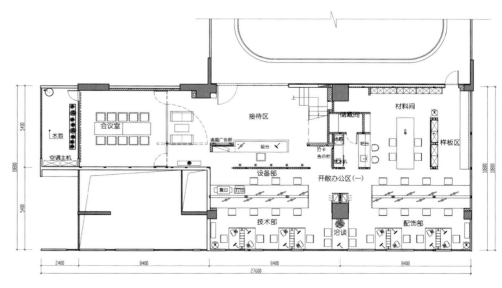

一层平面布置图

Because of business increase and scale expansion, the original office fails to meet the demand for company development in term of scale or image, and therefore a new office is established at the Xiangnian Square, Qiaoxiang Rd.

The original structure is a 4.8m high space, whose clear height is 4.2m exclusive of the beam height. How to design an interlayer without showing suppression in such a structure is a problem we need to solve. Based on analysis and research, pure white color is selected as dominant hue of the space; with the addition of reasonable space layout and accurate scale consideration, a clean and open space is shaped out with sense of structure and line.

Colors

The white color as dominant hue of the whole space not only demarcates a simple and elegant space atmosphere but also increases indoor brightness, thus making the space open and bright and bringing people joyful space feeling. The application of grey self-leveling floor, black door frame and stair steps breaks the paleness and wateriness shown by the much white color, make the space atmosphere vigorous and strengthen the sense of space structure.

Structure

Simple and smooth lines and ultra-thin floor slabs and steel plates outline the simplicity and elegancy of the space in aspect of structure and also deepen the space to some extent. The linear sense brought by the stair steps becomes part of space form. The bookshelf is made of ultra-thin steel plates covered with

white paint; it shapes out a simple model and echoes perforated wall windows separating book area from design area. Office desks and chairs, computers and lamps and lanterns are selected still based on the whole space structure.

Space Layout

The interlayer shows reasonable space planning. Vacancy above hall and office area is one of features of the whole space. Through usable space area is decreased, such design together with application of transparent glass expands the internal space depth and connects the external sights. And for employees, a more wide visual space is available and communication to each other is facilitated upon watching upstairs and downstairs through the vacant area.

The 90-degree rotatable wall between first-floor hall and conference room is another surprise in space layout. When the wall is set vertical to reception desk, an independent conference area will be segmented out; when it is set parallel to the desk outside of conference time, the hall and the conference room will be connected and then the outdoor sights will be introduced through floor glass in the conference room, making the space more open.

Lights

Ceiling lamp and wall lamp as main lighting facilities of the space are part of the space form. And the frequent use of spot light strengthens the expressive force of space. Lamp bands are embedded in the bottom of hall reception desk, background wall foot and side, thus solving the lighting problem incurred due to layer height and coinciding with the overall design language.

The design of new office is not only a representation of the moderate elegant design concept held by us for all the time but also a practice of design methods concluded by us. Not limited within specific design styles, all designs come from the rational analysis of site structure and actual demand, thus creating out a unique and targeted work.

Smarter manufacturing —— Top-class office integration

盘古投资办公室
Pangu Investment Office

设计单位：厦门喜玛拉雅设计装修有限公司
设 计 师：胡若愚
项目地点：福建省厦门市
建筑面积：3000 m²
主要材料：橡木、银豹大理石、手工羊毛地毯、
　　　　　灰镜、灰镜钢、银白龙大理石
摄 影 师：申强

Smarter manufacturing —— Top-class office integration

办公室设计既要凸显业主方的雄厚实力，更需彰显其大度从容的儒商风范。设计上采用现代东方风格，简洁沉稳又儒雅大气。空间上层层推进，通过不断收合，营造层次感和节奏感；从前厅、接待厅、书法长廊，四库全书休息厅、陶瓷藏品厅、董事长门厅，环环相扣，如宫院般森严秩序；而每个空间的主题营造，又凸显高雅文化品位。设计亮点：电梯轿厢门用不规则折面的木纹透光云石组合，电梯门一开启就已惊艳，而石墙上切割出的圆柱浑然天成，厚实有力，接待厅墙面如中式屏风的折面造型富有韵律感，反映在灰镜钢板的天花上幻影摇曳；中式祥云图样在手工羊毛毯上漂浮流动，而玻璃隔断上的山纹影像景随步移，不断分合变化，意境无穷……

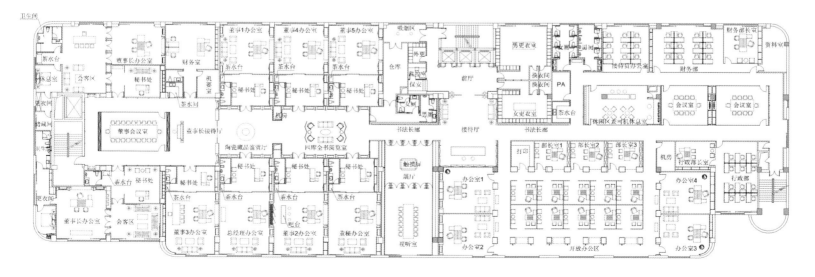

The office shall be designed to show the owner's power as well as its calm and elegant sense. It is adopted modern oriental style in design, which is simple but calm and great. The space has a sense of rhythm through continuous combination. From vestibule, reception hall, calligraphy gallery, Sikuquanshu Lounge, ceramic collection hall, to Chairman foyer, they are all connected with each other, like the palace structure. Each space theme is elegant and full of quality. The design highlights: the gate of elevator adopts wooden-texture marble with irregular folded surface. The moment you open the elevator door, you will be surprised by its charming. While the cylinder carved on stone wall seems naturally and powerful. The wall of reception hall is like Chinese folding screen that full of rhythm. It is dreamy when they are reflecting on the gray mirror ceiling. Chinese Xiangyun pattern is floating on the handmade woolen blankets. The image on the glass partition is moving with you, changable and meaningful.

Smarter manufacturing — Top-class office integration

Smarter manufacturing —— Top-class office integration

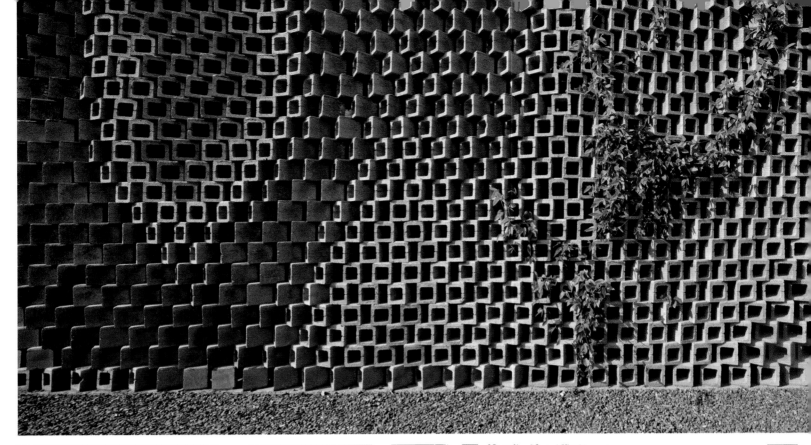

上海创盟国际建筑设计有限公司新办公室
New Office Space of Archi-Union Architecture Design INC. (Shanghai)

设计单位：上海创盟国际建筑设计有限公司
设 计 师：袁烽
建筑设计：韩力
室内设计：何福孜
结　　构：梅振东
给 排 水：李凤英
电　　气：潘吟宇
项目地点：上海市杨浦区军工路1436号
摄 影 师：沈忠海

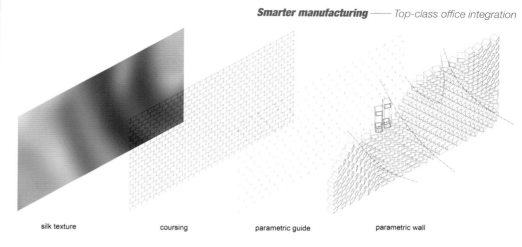

silk texture　　　coursing　　　parametric guide　　　parametric wall

Smarter manufacturing —— *Top-class office integration*

　　当我们面对废弃的第五化纤厂厂区中一栋三跨建于20世纪60年代的老厂房，并力图将其改造为创意产业办公空间时，寻找一种朴素的建造美学及一种真实而简单的建造过程成为我们的出发点。质朴的后工业景观和内在朝气勃勃的涌动，是我们对基地的感受。艺术创业者热衷于工业遗产中鳞次栉比的北向天窗、高耸的红砖烟囱、斑驳生锈的钢铁屋架。纷至沓来的婚纱摄影团队带着对对新人到此取景，渐渐构成了此地既流动又恒定的日常事件。前者来自过往，后者来自现实及未来。设计表达了我们的感受和场地的特质，而最能向大众清晰、直接地表达言语的介质应该就是最初设想的建构本质。

　　在整个三跨厂房的空间规划中，我们拆除了质量并不理想的第二跨厂房的屋顶。并将办公空间的入口布置在正对着第二跨厂房轴线的位置。这样步入整个空间感受到的首先是系列屋架下的草坪。这种无形的"旷"感，并非由卓越的形式所塑造，而是来自于简单的空间取舍。所有的墙体几乎保留了原初60年代的砖头，以求回归原始的与历史的美感。新种埴的爬山虎会在两年后将整个内院穿上新的绿色衣裳。整个空间设计的构思是通过空间整合与改造，塑造一种新的创意办公的生活方式。三跨厂房与原有的内院，经被拆除的第二跨空间调整之后，变成了由两进院子构成的系列空间。

　　外墙的设计出发点立足于真实的材料与建造表现。首先，是材料的真实性，我们拒绝虚假的粉饰与装饰性的非真实构造。采用的是最便宜的空心混凝土砌块体，力图从这种平实的基础材料中创造新建筑灵魂。设计全过程严格遵从建

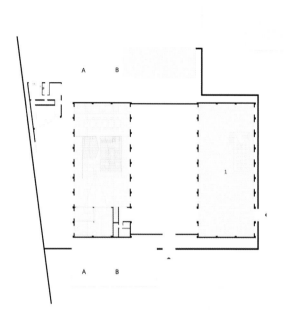

Ground Floor

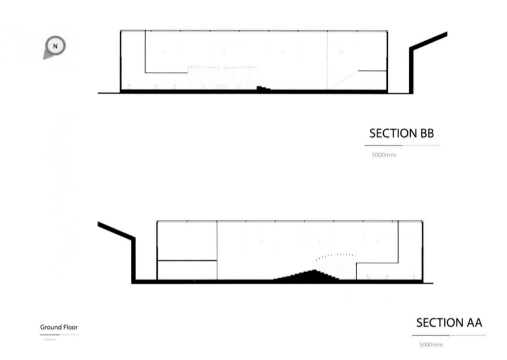

SECTION BB

SECTION AA

构精神，墙体材料采用的是空心混凝土砌块，但结构逻辑却非简单的砌筑。我们先行建立一个混凝土框架结构作为支撑体，在其外部完成混凝土砌块的砌筑工作。技术的运用基于我们对场地的把握和反应。面对曾经每日生产大批棉纺织品的场地，我们希望墙体能有织物一般的柔软质感，表达的内容具有丝绸质感的丝缎效果，这也是对过往的纪念。因此对丝绸质感的选择，成为思考的出发点。而对丝绸质感中灰度的读取则成为将丝绸转译为墙体砌筑方式的媒介。砌筑的砌块两侧通透、两侧实面，旋转角度不同，产生的阴影和视觉通透效果也不同，通过渐变的又有一定规律的旋转角度，使得墙体呈现出如织物般柔软、皱褶的效果。特别是远观时，效果更加明显。随着观看角度的变化、阳光的变化，皱褶状态也随之改变。

适当的妥协和变通使我们也体会到参数化给设计带来的好处，其中最直接的便是形式的更新——一种新颖的、但又不是完全与过往毫无联系的形式。

在施工一个月后，"雏形初现"的墙体便吸引了来园区拍摄婚纱照的团队和新婚夫妇，他们纷纷将其作为拍照背景。渐渐地，这堵墙便成为每位新人和摄影师的必选拍摄景点之一，我们设想这将成为此地的一种习俗。唯有形式的墙体只是一堵形式被固化的砌体块，但由新形式而促发的事件和不断流动的人群让墙体拥有了类似于生命体的更新。这是在意料之外的，但也是现阶段我们所做的最有成就感的事情——墙体将出现在越来越多的新人照片和他们的记忆中。

"绸墙"寄托了我们的城市乌托邦思想，这个实践的快速实现已指引我们走向"异托邦"，参数化的思维方式与低技建造的有机结合成为富有中国特色的参数化实践手段。期间，随意而出的是我们对传统文化的思考与内心解读。沉浸多年的纸面建筑在参数化的推动下，最终一定会回归到诗意建造的本质上来。而中国特殊国情下的实践应可在另一侧面推动建筑师对建筑设计与建造逻辑的重新定义和深度思考。

Facing an old three-span workshop built in 1960s in the abandoned Fifth Chemical Fiber Factory and trying to transform it into a creative industrial office space, we makes a start from looking for simple construction aesthetics and real and simple construction course. We feel that the base takes on a kind of modest post-industrial view and owns vigorousness inside. Art entrepreneurs are full of enthusiasm about skylight northward side by side, high red brick chimney, and rusted steel roof truss, in the industrial heritages. Wedding camera teams together with new couples take a photo here, which gradually becomes a flowing and constant daily thing. The former comes from the past while the latter comes from the reality and future. The design expresses our feelings and the site peculiarity. And the media to make a clear and direct expression in the best way is the initial construction nature assumed.

In the space planning of whole three-span workshop, we tear down the roof of second-span workshop without ideal quality. And we will set the office entrance at the position against the axis of second-span workshop. Thus, the grass under serial roof trusses will be felt first upon stepping into the space. Such intangible "openness" comes from simple space options instead of distinguished forms. The original bricks in 1960s are kept down almost for all walls, managing to get the original and historical aesthetic feeling back. Newly planted creeper will grow to cover the whole inner yard two years later. The whole space design is based on spatial inte-

gration and transformation, aiming to shape out a new life style of creative working. Three-span workshop and the original inner yard, after tearing down of the second-span space, are transformed into a serial space consisting of two yards.

The design of outer wall is based on real materials and construction performance. Beginning with real materials, we reject false prettifying and unreal decorative structure. Instead, the cheapest hollow concrete blocks are applied, striving to create new architectural soul out from such plain basic materials. With the whole design course in strict accordance with the construction spirit, hollow concrete blocks are applied as wall materials while structural logic is not just to build. We first set up a concrete frame structure as a support and complete building of concrete blocks outside. The technology is applied based on our cognition and reaction to the site. Faced with the site where a large amount of cotton textiles were once produced daily, we want to make the wall feel like soft textiles and show silk effect, which is a memorial to the past. Therefore, the silk texture is selected as the starting point of thinking. And the understanding of gray scale in the silk texture serves as a media to translate the silk into a way to build walls. Blocks are built with two transparent sides and two concrete sides as well as different rotation angles, leading to different shadows and visual effects. With the gradually and regularly changing rotation, the walls feel like textures, taking on soft and folded effect, which becomes more obvious with a long distance. And the folding state will change with the viewing angle and sunshine.

With appropriate compromise and flexibleness, we realize the benefits brought by parameterization to the design, and the most direct one is updating of form – new but in certain connection with the past.

After one month of construction, the wall "taking an embryonic form" attracts wedding camera teams and new couples taking photos here and is applied by them as the picture background. Gradually, this wall becomes a must for every couple and camera man, which, we assume, will become a kind of custom here. A wall only with form is just solidified blocks, but it becomes updated like a life when its new form inspires incidents and flowing crowd. This is unexpected, however, we are most proud of it at current stage – the wall will appear in pictures and memories of more and more new couples.

The "silk wall" contains our idea on city Utopia and its rapid realization has guided us to the "unusual Utopia". The combination of parameterized thinking mode and construction based on low technology becomes a parameterized practice means with Chinese characteristics. During the period, what is born from time to time is our thinking and understanding of the traditional culture. Pushed by parameterization, paper building immersed for years will finally return to the essence of poetic construction. At the other hand, the practice under special circumstances in China will help architects redefine and further think the architectural design and construction logic.

答案之门
The Way Out

设计单位：设计集人 (www.designsystems.com.hk)
设　计　师：林伟明、梁芬华、杨励莹、王永健、张芷茵、方欢欢
项目地点：香港
建筑面积：**238 m²**
主要材料：镀锌铁板、原有饰面铁板、粉末喷涂压坑铝板、中纤板

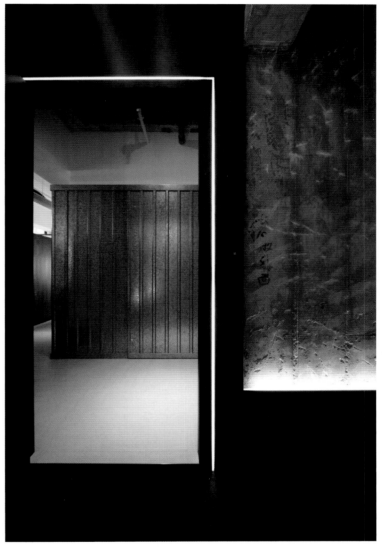

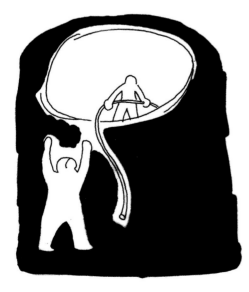

先从门口说起，大门门框嵌有一条灯槽，喻示了这束光线会为你带来问题的答案。进入办公室后，一条狭长的走廊作为会议室和工作室之间的过渡区域。阳光照亮了走廊的两端，引领你至办公区的方向。与阴暗的大堂及走廊形成对比，工作区域则营造出一片柔和、简洁和明亮的氛围，象征着客户在得到这间公司的帮助后成功解决问题的光明前景。

办公室内的墙体，工作台和灯槽都是为这项目而特别设计的。隔墙和灯槽的造型参考建造及工程上常用的波浪形金属坑板，作为反映这公司的性质与建造业的关系。除外形外，墙体特别做出实体墙、半通透及通透的效果以配合不同房间的私隐度要求。还有，锯齿形的灯罩能反射出更平均及柔和的灯光效果，有利于员工在台上使用电脑或工作。总的来说，我们对这些设计细节的重视，是想让人们产生一种如寻到宝物时那种"发现的乐趣"。这种对于细节的执着，亦能反映出这空间的使用者作为专业的顾问，能为其客户提供更高层次的专业服务。

Regarding the client company's mission to provide consultancy services on underground utility detection, the design aims to mirror the special uniqueness of the business nature by employing fashion-wise and custom-based design solutions.

The special business story begins with an elegant main entrance. The lighting design of the door frame vividly implies a different vision behind the door.

Contrast sometimes presents inspirational creation. The roughness of one wall contrasts with the clean and smooth texture on the other, bringing out a cool industrial feel. The dark wooden flooring, coupled with the white light up from the exposed overhead, appears functional while modern. Besides, the dark skirting is well teamed with the bright light from behind.

Right behind the door lays an elongated corridor to the meeting rooms and working area. Shiny sunlight creates a sense of direction to the inner space, giving the working area a warm, brisk and bright air. The amusing contrast between the dimmed entrance lobby and corridor conveys the underlying design metaphor of an alluring prospect ahead for this company's clients as they come here for a solution.

All partitions, working desks and the lighting are custom-designed for this project. The profile of the partitions and light reflector were developed from the corrugated metal sheet that is commonly used in construction business in order to give a hint of the company's business nature. These special elements were designed to suit different purposes: the partitions were constructed to be solid, semi-transparent and see-through to fulfill different degree of privacy required; and the zigzag shaped light reflector generates a more even and comfortable illumination on the desktop. Such attention to details provokes a sense of discovery in the viewers, as the same time reflects the professionalism of the client company.

Smarter manufacturing — Top-class office integration

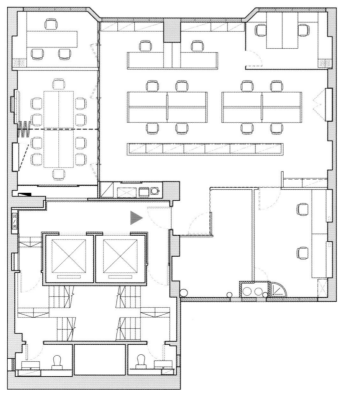

MARYLING服装中国区总部
MARYLING Clothes China Office

设计单位：于强室内设计师事务所
设 计 师：于强
项目地点：中国深圳福田区东海国际中心
建筑面积：**695 m²**
主要材料：高密度板白色哑光漆、白色/灰色乳胶漆、办公地毯、白色磁性手写板、玻璃

Smarter manufacturing — *Top-class office integration*

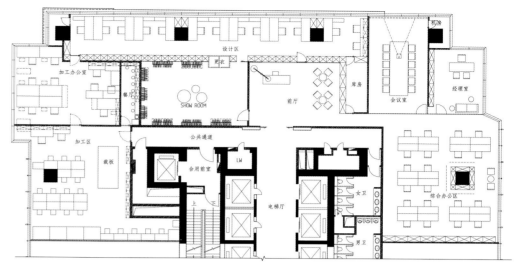

MARYLING是一家来自意大利的女装品牌,风格清新简约。作为中国地区的办公总部,希望延续品牌自身的特点,体现时尚的气息。

在室内设计中,MARYLING品牌特有的斜线组合图案,出现在大部分空间的立面上,包括一些功能性的储藏柜的立面设计也延续了那种斜线的分割。黑白的基调配搭部分红、黄等鲜亮的色彩,以及设计感强的家私体现了时尚的工作特性。

材料的应用方面,使用了很多简单机理的材质,如亮丽的编织地毯、橡木、白色亮光烤漆等。

MARYLING is an Italy female cloth brand. Its style is clearly and simple. As the headquarter in China area, it is hoped to extend its character to show the modern feeling.

In the interior design, MARYLING has its own diagonal mark picture, which appears in large space, including some storage cabinet surface. The black and white tone, coupled with red and yellow color, as well as the excellent furniture, they have shown the modern feature.

In terms of utilizing materials, it has adopted many simple texture materials, such as shining carpet, oaken and white shinning paint and so on.

Smarter manufacturing —— Top-class office integration

203

Smarter manufacturing —— Top-class office integration

腾讯科技(第三极)办公楼
Office Building of Tencent Technology (the Third Pole)

设计单位：北京艾迪尔建筑装饰工程有限公司
设 计 师：罗劲
参与设计：张晓亮、黄丽元、张清、董欣亮
项目地点：北京第三极大厦7层~11层
建筑面积：**21 330 m²**
主要材料：自流平地面、石膏板、钢化玻璃
摄 影 师：高寒

Smarter manufacturing —— Top-class office integration

腾讯科技一直信任艾迪尔,能够实现他们对办公空间的设计需求——活泼、和谐、创新,追求卓越。本次北京第三极办公室的项目设计主题是"最美的宇宙奇观",关键词是"神秘、无限、美丽"。我们分别用"银河"、"陨石"、"星座"、"彗星"、"流星雨"、"极光"、"星云"的具体概念塑造了"神秘的"各层办公空间。色彩上,我们不但延续使用了以往项目中常用的腾讯logo色系中艳丽的黄色和橙色,还增加使用蓝色、紫色来表现宇宙主题。整体风格现代又不失稳重,活跃又不失雅致。

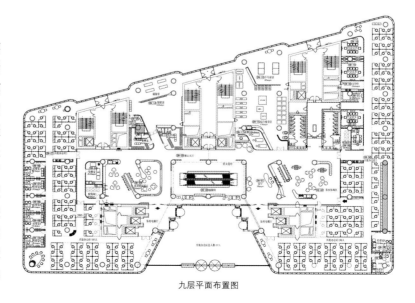

九层平面布置图

Tencent Technology has been trust IDEAL to achieve their design requirements for office space--lively, harmony, innovation, pursuit of excellent. The third class office building project in Beijing takes design theme of "the most beautiful universe spectacle". The key words are "mysterious, infinite and beautiful". We adopt "galaxy", "meteorite", "constellation", "comets", "meteor rain", "aurora", "nebula" to shape the "myterious" office space. In terms of color, we have continued the bright yellow and orange color used by previous projects of Tencent Technology logo. In addition, we have added blue, purple color to show the universe theme. The whole style is modern and calm, active and elegant.

十一层平面布置图

Smarter manufacturing —— Top-class office integration

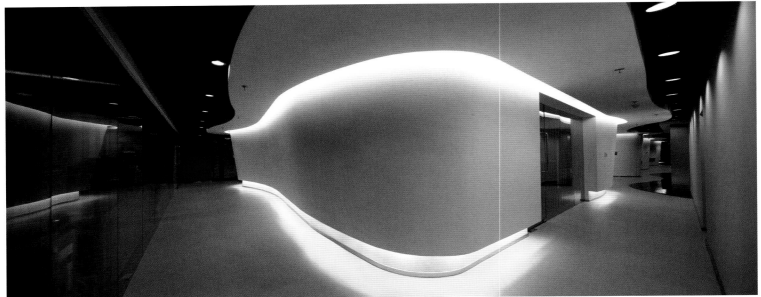

玲珑
Ling Long

设计单位：C&C(联旭)室内设计有限公司
设 计 师：吴联旭
项目地点：福建省福州市鼓楼区华林路246鸿源天城号B座8层306
建筑面积：180 m²
主要材料：科技木面板、大理石、木纹砖、蒙托漆
摄 影 师：吴永长

　　在有限的空间里奢侈地设计了一条连接室外入口和前厅的通道。设计十分大胆,狭长的通道,除了尽头两处灯光,没有任何光源,加上深灰色的地毯,整个通道充斥着静谧、幽邃之感,而轻舞罗袖的雕塑,肌理分明的雪松木饰墙,弥漫的木香,又增添几许神秘。设计的灵感来源于"道","道即本源",走在通道,让人忘记市井的喧嚣,静静地体会生命的意义,思索生活的本源。行数十步,豁然开朗,感应木门自动开启,灯光倾泄而出,映入眼帘的是前台灿烂的笑容,抬头处一片金黄的麦田。静谧和热烈、幽邃和明亮相隔于一门,此间道理道可道,非常道。

　　走进办公室,"静"和"净"是设计的主调,设计师抵制封闭单元格空间和接待与私密办公区清晰分隔的手法,韵律和渐变的概念得到发展,而功能性和空间数量并未受到影响。用简洁线条构成的几何图案将前厅、读图区、总监办公区、设计部巧妙地分隔,木作隔断又延伸了空间的层次。圆形的灯具柔和了线条的生硬。色彩上没有强烈的明暗对比,纯净白色与温润橡木色是空间的主色系。前台米白色系的伯爵石,墙面银白色系带着光泽反射的蒙托漆,天花纯白色色系的乳胶漆,不同白被赋予不同的表情层次,以不同分割造型予线条以立体,增添光影变化的可能性。也让"白色"在精心铺陈下有更多的想象空间。书架、办公桌、会客桌的简洁线条和落地灯的曲线、玻璃门上木作插销的圆形有机结合,律动了整个空间。工作之余,坐在飘窗前休闲区,抑或看看散落在窗台的闲书,抑或望望窗外的蓝天,抑或品味西河边见"上善觉"的禅意……生活是如此惬意,此中有真意,欲辩已忘言……

In a limited space to design a luxury outdoor entrance and lobby to connect the channel, a very bold design. a dark narrow channel, only two lights in the end, with dark gray carpet, the entire channel filled with quiet, secluded feeling , the dancing elegant sculpture, texture clear cedar trim wall, filled with wood, but also adds dash of mystery. Design inspiration comes from the "Road", "Road that is the origin of world, walking in the channel, people forget the hustle and bustle of the streets , quietly feeling the meaning of life, thinking about the origin of life.

Walking few steps, suddenly, sensor doors automatically open , light spilling out of the eye is the smile from reception, looked up at a golden wheat field. Quiet and warm, quiet, and then separated in a bright, it is a simple smart design and extremely clever.

Going Into the office, "static" and "net" is the main theme of the design, designers resist the closure of the cell space and reception area with a clear separation of private office practice, the concept of rhythm and gradient have been developed, and the amount of space and functionality and not be affected. With simple geometric patterns composed of lines make lobby, interpret blueprints District, director of office, the Ministry of cleverly designed to separate, wooden partitions and extend the space level. Circular lighting softened lines blunt. There is no strong light and dark color contrast, pure white with warm oak color space is the main color. Off-white color from the front of the Earl of stone, wall color with a shiny reflective silver paint Monto, ceiling pure white color of the paint, white is different given different expression levels in different lines to be split to form three-dimensional, added the possibility of changes of light and shadow. But also to "white" in the carefully lay out under the more room for imagination. Bookshelf, desk, reception desk clean lines and curve floor lamp, glass doors on the round wooden bolt combination, the rhythm of the entire space. Outside of work, sitting in front windows of leisure areas, or scattered in the window to see the light readings, or looked out the window of the blue sky, or the taste of the West River to see "the good sense" of Zen …… life is so comfortable, this there are true intentions, want to argue but forgotten words.

厦门共想设计办公室
Office of Gongxiang Design

设计单位：厦门共想装饰设计工程有限公司
设 计 师：郑传露
参与设计：庄钰麟、朱鹭欣
项目地点：中国福建厦门大学路沙坡尾58号
项目面积：400 m²

Smarter manufacturing —— Top-class office integration

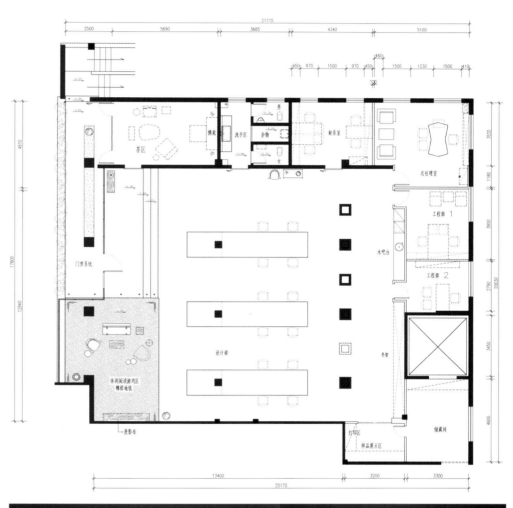

"作者"是一支充满激情,且热爱自然的设计团队,本次新办公室选址与"万国建筑博览"之岛的鼓浪屿隔海相望。本案开放的办公空间里,应用黑色的线条来强化空间的结构,废旧的木头、朴实的砖墙、蔚蓝的窗外,舒心且宁静!都说设计来源于自然,来自于对生活的理解,本案中设计师带着这种特有的心态去感悟每一个思维的创造。

"Author" is a passion and love for nature design team, this new office site and the "International Architecture Exhibition," the Gulangyu Island across the sea. Case open office space, application of black lines to strengthen the structure of space, waste wood, plain brick walls, blue windows, comfortable and quiet! Said the design came from nature, from the understanding of life! Designer case with this unique state of mind to create the perception of each thought.

Smarter manufacturing —— Top-class office integration

Smarter manufacturing —— Top-class office integration

朗诺经贸办公室
Office of Langnuo Economic and Trading

设计单位：**C&C(联旭)室内设计有限公司**
设 计 师：吴联旭
项目地点：福建省福州
建筑面积：**230 m²**
主要材料：橡木、大理石、地毯
摄 影 师：吴永长
设计时间：**2011**

　　白色与原木色，两种不同质感的色彩被鲜明地铺洒在这个办公室间内。它们没有过多的忸怩，但又不是直坦坦的，以致于我们很难一眼看透。这种感觉来源其实很简单，所以我们没有必要对着空间刻意找出它的原形。只要感觉到空间带来的新意，心情随之变得有些舒坦，还有些超越，那就好了。

　　在这个空间里，设计师将每个结构的关系表现得张弛有度。空间中没有刻意的渲染，也没有繁缛的修饰，一切看上去都简简单单。这种"简单"以开放的布局展示出最本质的生活内涵，它产生于材料所塑造的关系之中，并最终将这些内容和谐完美地融为一体。因为在设计师心中，简单不是最终的目的，而是在设计过程中寻觅本质内涵时所呈现出的状态。所以简单并不是单纯简约的问题，而是一个凝练的过程，是基于对保留和删减内容的把握。

　　简洁明快的线条，有种直截了当的美，让人们在面对一堵简单的墙的时候，也能回味着素面带来的快感。置身在这样一个素雅的环境中，它就象一幅随时会触发诗意的画布，让人满怀期待。在木质、白色瓷砖、人造石为主要用材的空间里，玻璃镜面适时地穿插其中，带给人们落落大方的亲切感。这种对空面的理解与再创作，将淡然和喜悦一阵阵地传递开来。

Whiter and wood color, two different colors are spreading in this office. They have not much self-conscious, but nor straight directly either. Therefore, it is difficult to see through them clearly. This feeling is very simple. Then, it is not necessary to find its prototype from the space. If there are come new ideas, and our feeling could be more comfortable, or even more, that's much better.

In this space, the designer allows the relationship between each structure to be suitable in terms of performance. There is no deliberately showing, or harassment on the modification, but everything seems to be simple but wonderful. Such "simple" are demonstrating the most essential meaning of the life in an open way. It comes from the relationship of the materials and merges these contents into a single whole in a perfect harmony way finally. Because in the mind of designer, simple is not the final purpose but the showing state during seeking the essential meanings. Then, simple is not only the pure and easy, but a coherence process. It is the skills on keeping or deleting the contents.

Simple lines, a kind of straightforward beauty, let people considering the nice brought by simple when facing such the simple wall. Standing such simple and elegant environment, it looks like a poetic canvas which will trigger your mind at any time, full of hope. Among the space taking wood and white tile, and artificial stone as the main materials, as well as the glass mirrors, they have brought a graceful feeling. Such understanding and re-creation of the space, the delight and joy are generated naturally.

Smarter manufacturing —— Top-class office integration

"亚邑"办公室
Office of "YAYI" Design

设计单位：上海亚邑室内设计有限公司
设 计 师：孙建亚
项目地点：上海闵行区
建筑面积：**350 m²**

Smarter manufacturing —— *Top-class office integration*

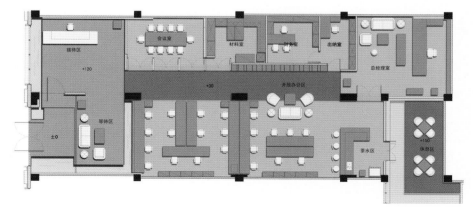

为"设计师"打造一个富有设计创意的工作空间,这本身是一个有趣并附有冲动的设计题材。设计师需要怎样的空间,才能更好地激发创作灵感?我们希望打破传统办公空间的有形格局,让空间"流淌"起来,为设计师提供一个能让思想自由生长的场所,创作"无间断,可发展"……

本案位于一个创意园区内,周边的绿化环境已经构成完好的生态空间,因此工作室的入口墙面采用大块面的玻

璃,尽可能弱化竖向阻隔,让空间通透,把绿化"导入"室内。工作室内部空间的处理也尽可能做到"无阻隔",开放式的办公环境有利于思维的活跃和交流。同时,我们尽可能选择环保、原生态的材料,尊重材质的本色,使用了廉价的青砖片,做为墙体的表现,吊顶裸露原混凝土结构。工厂直接出品厚10mm铁板,不经修饰打磨喷涂,作为门框及书架框结构。水泥石英砂地坪,及使用了拆房老木头地板。不刻意做旧,做到了"用旧如旧",永久不须翻新更换材料。于是,整个空间以最质朴的方式呈现,摈弃过多的人工痕迹,还空间以本来面目……这正是我们对于可持续舒适空间的理解和诠释。

我们一直相信,真正的绿色,不止满足于感官体验,更重要的是一种绿色的态度,它能够引导在这个空间里活动的人,向着更为可持续的生活方式发展,并同时影响着他们所服务的人群。

To create an innovative work studio for the designers is an interesting, refreshing, inspiring and challengeable topic to pursue.

What kind of space is the one may completely inspire and motivate the designers for the fabulous and refreshing designs?

We opine that to have the designers gotten out of the traditional layout of the office building and the relationship between the human being and the building, to provide the designers a cross-functional and boundless space that it may bring them the solid inspiration with sustainable development…

The Project is located in a Cultural Creative Community with a well-designed ecological landscaping, which is surrounded by trees and plants.

Curtain wall facade is recommended for the main entrance of the workplace, allowing continuity of outside environment penetrating into interior space. Internally, working spaces should not be separated as much as possible.

Open planning office environment enhances creativity stimulation and interaction.

For selection of building materi-

als, we suggest to pick up those with natural appearance, environmental friendly finishes, respecting its original texture and color, e.g. Fairface concrete, natural wood, stone, etc. can integrate harmony in the same space. This creates a representation of the original character of materials, minimize artificial articulation.

We insist and believe that the real safe and environment-friendly of the "Green" is the one can not only provide people with the comfortable feeling, but also make people completed understand what is the attitude of environment-friendly of the "Green", which can lead a life of sustainable development for the human being who have activities in the Project, and also affect the people who are served by it.

杭州意内雅办公室
Yineiya Space Design for Offices in Hangzhou

设计单位：杭州意内雅建筑装饰设计有限公司
设 计 师：朱晓鸣
参与设计：赵肖杭、高力勇、侯兴善
项目地点：杭州市西岸国际艺术区 11 号楼 B 座
建筑面积：750 m²
主要材料：钢板腐蚀、回购旧地板、人造石、水泥、
高密度板、FC板、地胶板
摄 影 师：陈乙、蔡刚

　　本案位于杭州京杭大运河拱宸桥西岸,为杭州运河改造五大工程之一的运河天地文化创意园内的一幢近代工业厂房遗址。建筑周边环境历史文化气息浓郁,悠长的京杭大运河似乎还在回忆那繁华的水运年代,园区内绿树成荫,几幢粉墙墨瓦的老宅院静寂地诉说它曾经的历史;"人去楼空"的近代厂房遗址暗示着这里原是运河工业文明的中心……在如此富含历史文脉的优越外围环境下,设计师在执笔设计前对建筑室内空间形态的畅想尤为强烈。该如何去对接周遭的环境?是新锐主张的割立,还是传承延续?如何表达怀揣理想的年轻创意群体的办公与生活方式?如何体现loft的本质意义?是改造还是创造?是千人一面的斑驳老墙,漫无目的的涂鸦?还是回归到创造loft最初是享受艺术的空间重组带来情绪的愉悦?

　　也许,往往简单的手法却能表达一些复杂的思想。卸下枷锁回到最初,回到最初人对loft空间的需求,表达自我,满足自我心理、视觉空间的需求。

　　经过现场的勘察与感受,几番斟酌,我们将该建筑分设为"抑"、"扬"两区来规划。将建筑内留舍原建筑12 m高度为中庭,围绕着中庭根据办公功能需求,割化了公共、过渡、办公、娱乐休息、辅用等空间。在中庭的设计中引用了极简化处理的院落概念,将不同性质的空间形成围合形式,即保留了良好通透的心理空间和视觉上高度的享受,又将自然光源均匀地引入办公各区域中。

　　整体中庭置于水景之上,鸟笼、树池

点缀其中，映射了运河水文化的同时，使得整体空间更为静谧与灵动，小有一番"鸟鸣林愈静"的意境。中庭前后分别为公共接待区与公共办公区，先抑后扬的空间表现手法，加强空间体块的错觉，过廊保留了原始的斑驳老墙与锈迹钢板，并赋予其几何形体的立体构成形式，显现了一种虚实结合的光影效果，及引导来访者产生好奇的情绪，移步中庭，场景豁然开朗。搭建成阶梯状的三层钢构办公区独立于建筑内部，"楼中楼"的建筑形体增添中庭"空"的视觉感受。在自然光源不足的区域中，恰当安排成制作间、材料间、茶水室等辅用空间，并根据年轻群体的行为特质，增设了影视、娱乐、用餐区域，劳逸结合提升工作情绪上的愉悦感和满意度。

为表现纯粹素雅又不失矛盾的空间印象，色彩中起用了对比强烈的黑白灰色系，材料的应用上大量使用回购的老木板、钢板、旧砖，并部分采用火烧碳化、腐蚀的方式加强材料肌理，也有现代的材料如自流平、FRP、地板胶、贴膜玻璃、镜面不锈钢等的运用，旨在于拉大材质间的对比反差。

整体空间风格中，注重中式借景手法却赋予其后现代的形体和材质机理，元素有矛盾也有包容，其本身呈现一种特立的风格。壁炉、青砖、帝国时期风格的欧式灯具、鸟笼、美式球……是中式？美式？后现代？说不清，道不明的风格关系也打包在了一起。

Loft是开放的、创新的、是包容的、矛盾的、多样的、流动的；是表达自我主张的建筑语言；是物质的，也是精神的！

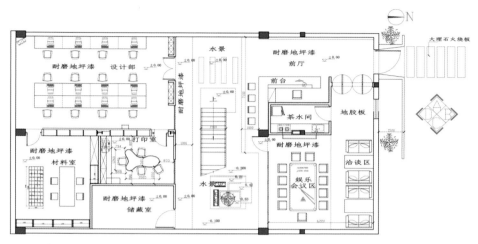

一层平面布置图

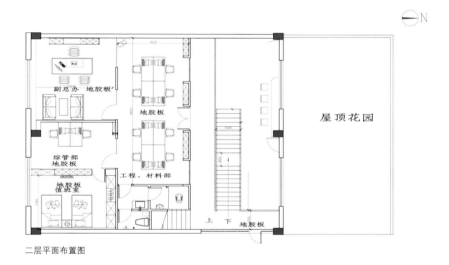

二层平面布置图

三层平面布置图

Located at the west bank of Gongchen Bridge over the Beijing-Hangzhou Canal in Hangzhou, this case refers to a modern industrial plant site in the Canal World Cultural Creative Park – one of five major reconstruction projects of canals in Hangzhou. The building is surrounded by strong historical and cultural atmosphere where the long Beijing-Hangzhou Canal seems to be recalling the flourishing canal year. In the park, trees can be seen everywhere and several old houses with pink walls and inky tiles are lonely and telling their histories; the modern plant site "full of loneliness" implies here was the center of industrial canal civilization…. Under such superior environment with rich historical context, designers cannot help developing their thoughts about the space shape of building inside before starting their design. How to make a connection with the environment around? Insist on a new and vigorous opinion or inherit the old one? How to show the working and living styles of the young creative groups with ideals? How to reflect the essential meaning of loft – make a transformation or creation? Focus on the old mottled wall with aimless doodles? Or go back to the initial loft creation when the artistic space reorganization brings joyfulness?

Sometimes, maybe, simple skills will always be able to express some complex ideas. Unload fetters and go back to the start – the demand for loft space, to express self and satisfy the demands for self psychology and visual space.

With survey and experience on site as well as consideration, we will plan the building based on "falling" and "rising" styles. A hall will be designed based on the original building with a height of 12m, around which spaces are separated for use of public, transition, working, entertainment and auxiliary according to office function needs. A simplified yard concept is applied in the hall design where spaces with different natures form an enclosure which not only remains a good and transparent psychological space and a high-level visual enjoyment but also makes the natural light enter office areas uniformly.

The whole hall lies on waterscape decorated by the birdcage and tree pool, which reflects canal water culture and meantime makes the whole space more quiet and magic, thus almost forming an artistic conception – "bird singing reflects forest stilling". The front and back of hall are respectively designed as public reception area and public working area. Such space design technique which can be called "rising after falling" strengthens the illusion on space blocks. The original old mottled wall and rusty steel plate remain in the corridor; their geometric forms are changed in a three-dimensional way, thus showing a shadow effect of combining blankness and actuality and making visitors curious who will become enlightened after they walk into the hall. The three-layer ladder-shape working area of steel structure is independent of the building inside. The "building within a building" form makes the hall visually felt more "hollow". The area without enough natural light are properly designed into production room, material room, rest room and other auxiliary rooms. Besides, according to behavior characteristics of young people, areas for videos, entertainment and meals are added, thus allowing combination of working and resting to promote joyfulness and satisfaction for works.

To make a space impression with pureness, simplicity, and contradiction, colors of black, white and grey with strong contrast are applied. As for materials, old wood boards, steel plates and old bricks are repurchased for use in large amount and part of them is subjected to carbon-

ization and corrosion for strengthening of textures; modern materials are also applied, such as self-leveling floor, FRP, floor glue, coated glass and mirror finished stainless steel, aiming to strengthen the contrast among materials.

As for the whole space style, scenery borrowing skill in Chinese style is applied and meanwhile post-modern mechanism on shapes and materials are added, thus forming elements with contradiction and inclusiveness, which shows a unique style. Fire place, grey brick, European lamp with a style of empire period, bird cage, American football⋯. All these things are of Chinese style, American style, or post-modern? The answer is uncertain and meantime the uncertain style relations are packed together.

Loft is open, innovated, inclusive, conflictive, multiple and flowing; it is a kind of architectural language to express opinions; it is material and also spiritual!

赢家商务中心
Beijing Winner International Business Service Co., Ltd.

设计单位：RVG Design & Solution Co.,Ltd
设 计 师：张海涛
项目地点：北京中关村银谷大厦
建筑面积：600 m²
主要材料：大花白云石、定制吸音板、烤漆玻璃、定制地毯、BOLON编织地毯、镜面不锈钢、高革硬包、染色小牛皮、杜邦可丽耐
摄 影 师：高寒
撰　　 文：张海涛

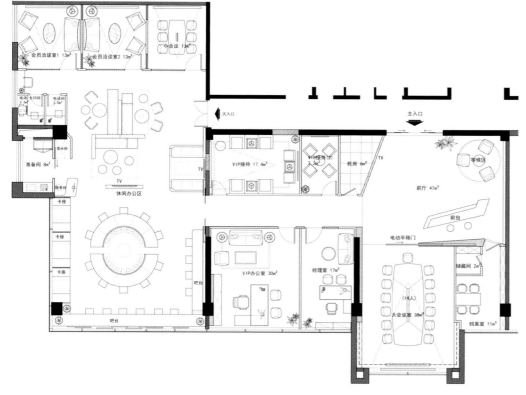

随着国家政策的鼓励和移动互联网技术的革新，在第一代互联网创业潮后，国内又迎来了新一轮的创业高峰期。而对于初期创业者来而言，都希望能寻找到一个既能满足形象需求又价格实惠的办公环境。针对这个市场需求，虽然国外一些知名的服务式办公环境提供商（譬如REGUS、SERVCORP等）较为成熟，但价格还是相对较高，一些资金不富裕的个人创业者还是很难接受。而RVG这次承接的项目委托方就是瞄准这样的市场定位，希望能为这样一群初期创业者提供一个高性价比的办公服务，充分满足他们的需求并希望能成为创业企业的一个孵化和成长基地。

RVG在接到这个项目委托之前，虽然做过大量的办公空间案例，但主要是针对直接用户需求的设计，却很少关注这种租赁服务项目的设计。因此我们特

地做了深入的调研和沟通,以便更准确地把握住最终用户的实际需求和使用体验。客户给出的租赁客户定位是30岁左右的互联网及高科技领域创业人群,我们的调查结果也和客户定位基本一致。因此,接下来的工作重点就是如何满足目标客户的使用习惯和审美要求了。

互联网及高科技行业的特征、基本都是80后的创业群体,"科技、时尚、交互、开放"成为我们设计时的关键词,同时我们还是觉得应该再加一些概念进去。对于小企业,快速和持续的成长是最好的一个状态,我们需要一个体现成长的元素,最终,我们在开放办公区种上了一棵"成长之树",以弯曲盘旋而上的白色钢管作为树干,放射性的同心圆灯带作为树蔓,地面对应的拼织地毯作为树根,喻示着事业在从里开始扎根、成长、开枝散叶。

在设计之初,委托方就向RVG表达了要充分提高利用率的要求,只有这样才能降低运营成本,以便以更优惠的价格提供给客户,在不到420 m²的套内面积内,要满足各种功能,大小会议、独立办公、洽谈、开放办公、休闲区、电话间、文印间、茶水点心间、设备、档案等等,各种功能一个也不能少,还要不能显得太拥挤。为此,在平面布局上的确推敲了很久,在通过几轮的调整后,最终还是得到了大家的一致认可。

作为一个创业孵化基地,需要满足不同客户的不同使用偏好,因此通过创意的规划和设计来带给他们新鲜的体验成了我们的主要任务。在开放区和休闲区我们根据平面的特点,设置了多种样式的办公区,有吧台式、圆形岛式工作台、临窗卡座式、半封闭包间式等等,RVG希望通过形式的不同持续地给使用者带来不同的体验,不同的人数或团队可以选择不同的办公方式,自由、舒适或许才是他们最想要的。由于本案位于较高的楼层,天气好的时候可以看到西山,也可以俯瞰整个中关村,冲一杯咖啡,晒着午后的阳光,找个临窗的座位,打开笔记本,很惬意地享受工作。为了呼应"成长之树"的概念,整体的开放办公区和休闲区都是在白色的基调上以柔和的草绿为主色调,绿色有春天的感觉,代表着生长的力量。

空间中另外处理的重点就是集成了最先进无线智能控制技术的多媒体大会议室,为了迎合目标客户的习惯偏好,RVG在整个项目中大量引用了较为先进的科技装备和技术措施,希望能为客户提供更便捷更人性的空间体验。为了增加前厅空间的采光和展示出多媒体大会议的空间,我们设置了一个我们迄今为止设计过的最大的单扇电动玻璃平移门,近3 m的宽度,开启时还是有点新鲜感的。在这个项目里我们也首次采用了国外最新款式的吸音板做法,在保证功能的时候兼顾了效果的美感,这是我们比较满意的一次尝试。多媒体大会议的视频、音响、灯光、场景等均是通过无线设备控制,很符合当下的无线办公趋势,很先进,RVG相信,在日后使用中,当创业者在这里给他们客户汇报时,这个会议室能为他们挣足面子。

为了倡导低碳和绿色,在整个项目中,85%的照明都采用了LED光源和技术,虽一次投入较多,但长期来看还是非常节能和环保的。希望能通过我们的努力,让创业者们能获得良好的办公体验,也祝他们的事业能快速成长!

With the encouragement of national policies and the technical innovation of mobile internet, a new peak time to start an undertaking comes after the first-generation internet tide. In consideration with such market positioning, RVG, the entrusted part of this project, hopes to provide the early entrepreneurs with office service of high performance-price ratio, which can fully meet their demands and become a base for breeding and growing of undertaking-starting enterprises.

Before the project is entrusted, RVG has completed a large number of cases about office space. However, most cases concern designs aiming at direct customer demand instead of designs of rental service projects. Therefore, we especially make an in-depth survey and communication to correctly know the actual demand and usage experience of the final user. The client mainly positions a group of people around 30 ages and engaged in the internet and high-tech fields, with which our survey results basically coincide. So, the next work shall focus on how to meet the use habits and aesthetic requirements of the target customers.

The internet and high-tech fields are almost filled with the generation after 80s. Therefore, we select "science and technology, fashion, mutuality and openness" as key words of the design; in addition, we will add some other concepts. For small enter-

prise, rapid and continuous growing is the best state. Then we need an element reflecting growing. Finally, we plant a "tree of growth" in the open office area. With spiral-up white steel pipe as trunk, radial concentric-circle lamp band as tendril and weaved floor rug as root, the tree implies that career is rooted inside, grows and then flourishes.

At the beginning of design, the entrusting party expresses its will that RVG shall make full use of the space to reduce operating cost and thus to offer clients a more preferential price. In a space with an area under 420m2, various functions are included, such as large and small conferences, independent working, conversation, open working, recreation area, phone room, printing room, resting room, equipment and files. The space shall not look too crowd when all these functions are realized. Therefore, the layout is considered for a long time. After several times of adjustment, everybody is satisfied finally.

The enterprise breeding base shall meet different preference of different clients. So, our main task is to bring them new experience through creative planning and design. Based on planar features, we design in public area and recreation area an office area of different styles, such as bar-counter style, round-island workbench, window booth style, and semi-closed compartment style. RVG hopes different styles can continuously bring users different experience. The working way can be selected based on different person number or team, in consideration that freedom and coziness maybe what they most want. This case is located at a high floor. In case of a sunny day, the Western Hills can be seen and the whole Zhongguancun can be over-

looked. Then, with the afternoon sunshine, you can make a cup of coffee, find a seat by the window, turn on the laptop and enjoy your work. To echo the concept of "tree of growth", the whole public working area and recreation area is designed with white as the basic hue and soft grass green as the dominant hue. Green brings people a feeling of spring and represents growth power.

Another key point in the space is the large multi-media conference room equipped with the most advanced wireless intelligent control technology. To cater for the habit and preference of target clients, RVG applies in the whole project a large number of advanced technical equipment and measures, hoping to offer clients more convenient and human space experience. To strengthen the day lighting in front hall and show the large multi-media conference room, we design a single-leaf electrically-operated glass translation door, which is the largest one we have designed so far. The door is near 3m wide and can bring some fresh feelings when it is opened. In this project, we apply for the first time the sound panel of latest foreign style. We are satisfied with the trying for it ensures both function and aesthetic feeling. Videos, stereo system, lightings, and scenes in the large multi-media conference room are all controlled by wireless devices, coinciding with current wireless working trend.

To advocate low carbon and environment protection, 85% of lightings in the project apply light source and technology concerning LED. In spite of much input for one time, such way is energy saving and environmental on the long time.

树下
Under the Tree

设计单位：竹工凡木设计研究室 / CHU-studio
（www.chu-s.com）
施工单位：竹工凡木联合工程团队
设 计 师：邵唯晏（Alfie Shao）
参与设计：邵方玙、林予嵂、许思敏
建筑形式：透天昔
房屋状况：新旧混搭
项目地点：台湾桃园大溪
建筑面积：**95 m²**
主要材料：钢构、木作、抛光石英砖、强化玻璃、亚克力、冷烤漆、碳化木
摄影师：竹工凡木设计研究室 / CHU-studio

　　本案位于大溪市区,业主夫妇对于3名子女的教育相当重视,除了翻修老屋外,更增设美术安亲教室,期待透过空间环境的色彩风格与明快设计感,借重鲜明意象在生活中启发子女的艺术感与创造力。因年轻业主夫妇两人均留学国外,对前卫观念的艺术品味与设计事物接受度高,因而特别要求空间特质的营造。对此,我们从整体意象思考空间表现的可能性,提出在空间置入向上生长的大树做为设计概念意象,以大树生长对照树下学习的意象,回应业主对于子女教育的强调与用心。再者,除艺术性外,"大树"同时拥有更重要的责任与功能。

结构考量

　　因一楼前段是传统房舍,在旧结构之上继续搭建转换为三层楼的透天昔,因所有楼梯挑空位于同一位置,加上当时建物新旧交接面处理不当,导致整体建物往一楼前段微倾,虽不致于影响建筑安全,但讨论后还是决议局部补强,树中结构以H钢补强,由下往上的绿色树干,长出地面蜿蜒攀附穿过楼层,向外蔓生高低依托墙面伸展而上,从垂直向度巧妙补强了挑空后的楼板与结构强度。

设计数位性

　　设计过程也是本案受业主青睐的重点,透过电脑演算的方式,给予空间几个结构考量下的相对座标,让"大树"在虚拟环境中计算,产生可能的支撑形式,后端再由设计师接手处理。天花部分以三

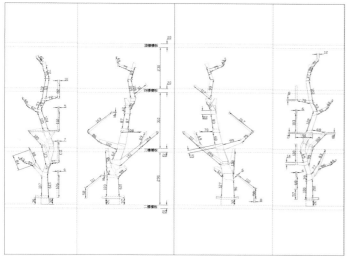
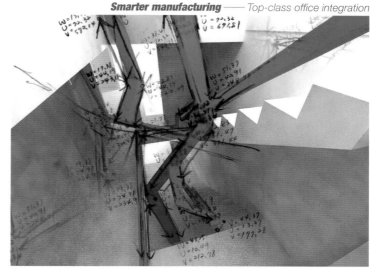
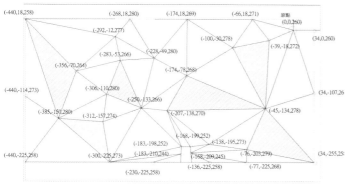
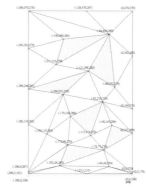

角形为原型进行数位演算,渐次将二维的三角形错落散布天花板,复以挑空中的大树为中心,运用精密的折板系统将天花板翻折成立体块面,由枝干伸展依附的墙面天花开始变化,沿着挑空周边建构出延展的起伏角面,企图营造动态树荫之感。而工程部分,透过镭射座标定位,配合电脑辅助系统(CAD/CAM),就能精准快速完成组立工程。

美术安亲教室

业主夫妇爱好艺术,找了几位老师教导美术创作,大树的意象就是为了营造一个小朋友的想像天地,借由抽象树的造型语汇,让空间蒙上一层梦想的氛围。所有房间及教室都围绕着大树,小朋友无论在图书室看书或是教室中上课,都能看到大树毅力不摇伫立在中央天井区。

作品展示

为回应美术安亲教室的需求,大树也提供许多放置小朋友作品的展示机能,透过吊挂及展示平台的方式,让小朋友的作品蔓延至整个空间,甚至可以在大树上涂鸦,将活动与记忆烙印在树上,同时大树也成为小朋友间最佳的留言板。

清透扶手,增大空间效果

本建筑为长向街屋,虽纵向够深,但短向过窄,再扣掉楼梯及房间,空间显得狭隘,因此我们保留建筑既有挑空处,并透过强化玻璃配合树枝意象,巧妙化身成为轻透的楼梯扶手,借此让空间有增大之感,同时鲜明绿色既是视觉焦点,也引导着上下穿越的行进动线。

透过建筑中央切出挑空,引入充足光线,解决多楼层常面临的光线不足问题,并创造随自然变化的空中光井。

本建筑为长向街屋,因而减少各层隔间墙,降低光线与视觉的阻碍,仅留下最少量的墙面界定空间分区,以此维持空间的开放、穿透与连续的整体感。

将隔间墙上端切除以清透玻璃处理,目的让各处都能见到大树,同时有光线均匀散布之效,给予小朋友明亮的学习环境。借由光线引入与视觉延伸的穿透性,各楼层空间平时流动相融,遇有不同活动需要各自遮蔽、分隔时,搭配轻薄门片的推拉开展,即可达到彼此空间界分的独立性格。

The case is located in Tahsi. The house owners quite emphasize on three children's education, so in addition to renovating the old building, they also established a Kids' art classroom within the house in the hope of inspiring their children's creativity and cultivating their sense of art. Since the young couple both have studied abroad, they pay much attention to the modern art and edgy design, and value the characteristics and ambiance of the space. We tried to find different possibilities in the space through the whole image and came up with an idea of installing a "giant tree" in the house, for trees in traditional are a source of strength and shelter, and a symbol of parents and their responsibility. Besides the artistry and imagery of the tree, it also has important functions in the house.

Structural

As the front of the first floor is a traditional Fujian-style house, we continue to build and renovate the old structure into three floors townhouse. Because the open areas of all stairs are the same place, plus the junction of the old and new buildings were not handled properly, it led to the tilt of the building. Though it did not affect the construction safety, we still decided to reinforce it by using H steel reinforcement. The green trunk growing from the bottom to the top and climbing the wall to the ceiling turned out smartly reinforce the structural strength vertically.

Digital design

Designing process is also the focus of the case favored by the owners. By using the computer operation process, we worked out the tree in a virtual environment, producing a possible supporting form, and then let the designer take over the back-end job.

Triangle as the prototype for the digital calculation of the ceiling and the tree as the center, we spread the 2D triangle gradually on the ceiling and use precise folded plate structure system to turn the ceiling into 3D. The changes of ceiling started from

the trunk and branches attached to the wall and stretched to ceiling, creating a natural shade of the tree. As to work of positioning, through the laser position system and CAD/CAM, the folded plates can be assembled and completed by the carpenters quickly and precisely.

Kids' art classroom

The couples love art and crafts, and they have employed some professional teachers to teach fine art in the classroom they established. The image of the tree creates a fancy world for kids and, with its abstract shape, the classroom is now full of imagination and learning atmosphere, because wherever in the classroom, every kid can see the tree and feel its accompany.

Function of display

As an art classroom, the tree also functions as an arts a works display for children. It is also a message board for children to make friend and express their thoughts.

Transparent handrails

The building is a long approaches street house. Though the long-side is deep enough, the short-side is just too short. As the space was quite narrow, we retained the existing high ceiling and installed strengthen glass, making the space look more spacious. In the meantime, the green tree has also become a visual spot and a route guide of the whole space.

The building brings sufficient light to every floor through the skylighted central area. It also solves the common problem of insufficient light, and creates a natural and changeable light source from atrium.

The building is a long approaches street house, thus we try to reduce the partitions and the obstacle of light and sight in order to make it look more spacious.

We removed the upper part of the wall with transparent glasses in order to keep the tree be seen everywhere. By doing this the light can be spread evenly in the house, and children can get a bright and suitable learning environment.

Through the flow of light and visual extension, each room and each floor are linked and connected. But when individuals need a little space, sliding doors provides the function of separating difference spaces and keeping privacy.

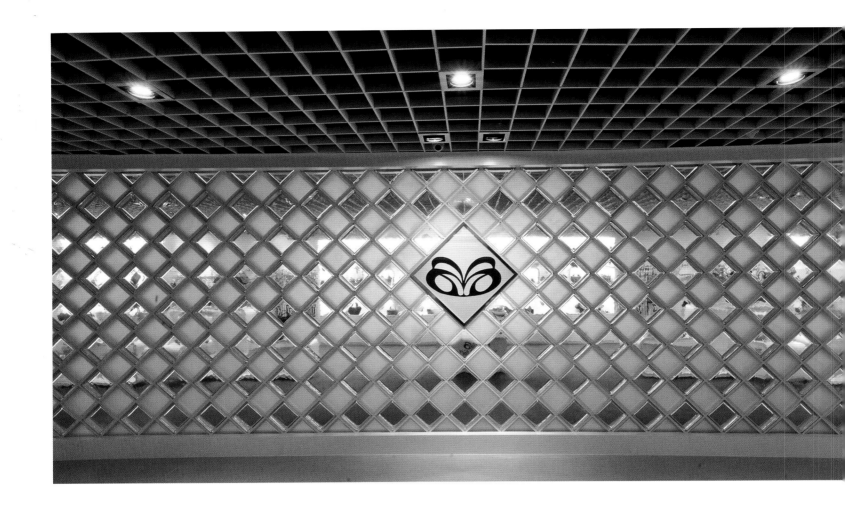

山二集团企业总部
YAMANI Group

设计单位：竹工凡木设计研究室 / CHU-studio
　　　　　（www.chu-s.com）
施工单位：竹工凡木联合工程团队
项目地点：台湾台中市
设 计 师：邵唯晏（Shao Wei-Yen）
参与设计：汪伯成、林其圻、邵方璵、林予帏、许思敏
建筑面积：**200 m²**
主要建材：石英砖、石材、玻璃砖、白铁变管、黑铁、**PVC**
　　　　　地砖、特制地毯、进口皮革、镭切亚克力
摄 影 师：竹工凡木设计研究室 / CHU-studio

Smarter manufacturing —— Top-class office integration

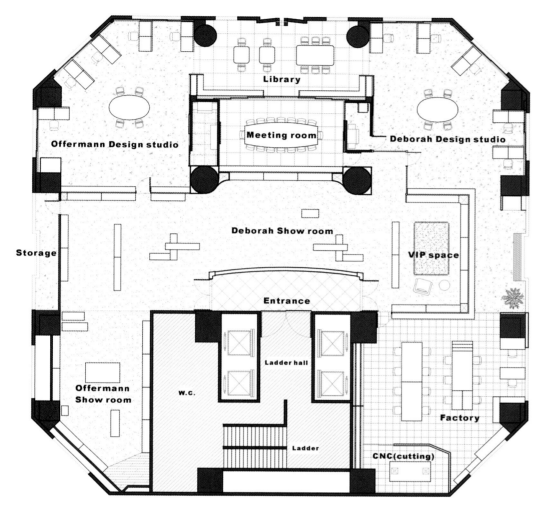

本案为山二集团位于台湾台中的总部设计规划案，山二集团早期为顶极皮包代工，进而转换跑道自创品牌，积极在台湾及大陆市场拓展设点。总部的前段为Deborah女包展场及Offermann男包展场，中段为阅览空间、接待空间及会议空间，后段为2个设计研发部门及一处打样制作部门。规划的主轴在于透过店中店的手法来强化展示的机能，借由低调奢华的质感融入当代设计语汇，让参观者进入本空间后能先感受到展场细腻的气质，同时也能参观部分R&D空间，感性与理性空间的铺陈为设计规划的主要诉求。

为营造感性与理性共存的空间氛围，我们将既有天花系统全部拆除，使用穿透感强烈的开放式格栅天花系统，在以白色及大地色系的简约质感下，隐约还能看到上方原始粗犷的天花系统。设计研发部也借由粗犷材质的使用，企图透过细腻质感与粗野味道的撞击来营造空间张力。再者，在所有展示柜的部分，我们利用模矩化的思维打破传统皮包展示柜体一成不变的形式及做法，透过几种规格的模矩可随不同的需要任意排列组合，便于适应未来拓展设点时的需求，同时也透过变与不变的形式组合来呼应企业品牌对于未来的展望和企图。

This is a design case for headquarter of Shaner Cooperation in Taiwan. In the beginning, Shaner Cooperation is a biggest bag foundry, and then it switched to create its own brand, expanding its market in Taiwan and the mainland. The first part of the headquarter is the exhibition of Deborah for ladies and Offermann for men. The middle part is reading space, reception space and meeting space. The last part consist with two Design and R&D departments and one sample making department. The axis of the plan is to show the functions through the shop in shop way. That is adopting a low-key extravagant to the contemporary design, which is to provide visitors a delicate temperament to te space and allow them to feel part of the R&D space. the emotional and rational space are the main factors of this design.

In order to create an emotional and rational coexistence space, we remove all the existing ceiling system, and adopting the open-grid ceiling system which has a strong sense of penetration. Under the white and

earth color as the background, we can see the orginal ceiling system slightly. The Design and R&D department has adopted the crude ore material to create a space tension through the delicate and rough sense.

In addition, for all the display cabinets, we has adopted the module thinking to break the traditional bad display cabinet. Through different modules, we can create different combinations according to the different purposes so as to meet the future expansion needs.

Meanwhile, through the changed and unchanged combinations, it is a echo to the spirits of this cooperation for the future brand.

设计部办公室
Ministry of Design Office

设计单位：**Ministry of Design**
设 计 师：**Colin Seah**
项目地点：新加坡
建筑面积：**345 m²**
设计时间：**2010**
摄 影 师：**CI&A Photography、Edward Hendricks**

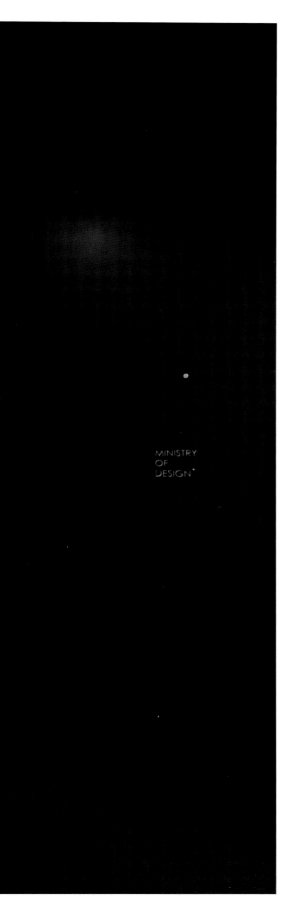

MINISTRY
OF
DESIGN*

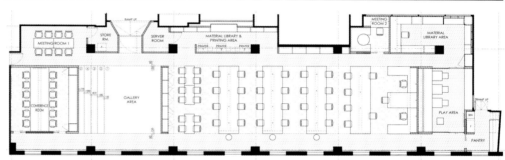

　　MOD的办公室由六间经过改造的店面构成,位于新加坡具有历史特色的唐人街与CBD交汇处。办公室秉承的核心原则是材料与色调的和谐以及"精炼"的概念,这一原则也指导着MOD的设计工作。

　　为了找到能实践创新想法的地方,我们决定寻找一个足够大的场所,可以将整个办公区容纳在同一楼层里。我们强烈感觉到开放的交流对于创新和创造至关重要,因此,我们的办公空间里将不会有任何的等级差异或者隔阂,而是一个真正自由的办公区。

　　工作室内双基准线的外沿间精心设计出了一些线性空间,使得环绕整个空间形成流通轴心。同时在终点处安装镜面,可以拉伸流通轴心,形成日常的"T台"。跟条形码相似,沿着这些"T台"可以分布不同的区块,从艺术区、会议区、创意区、议案讨论区到图书室。

　　所有外来的都被当做已有空间里的物体,但在在视觉上还是保持他们的独立。整个艺术空间可以不断更新变化,这样既可以有一个表达自我的平台,又不需要对整个工作室进行重新设置。我们刻意将工作环境设计成单纯的黑白色调,这样才能进行创造性的探索而避免中庸。图书室采用的是半加工的木料,从构造学上来说,通过与周围空间形成对比效果,它可以创造出一个更加放松的环境。

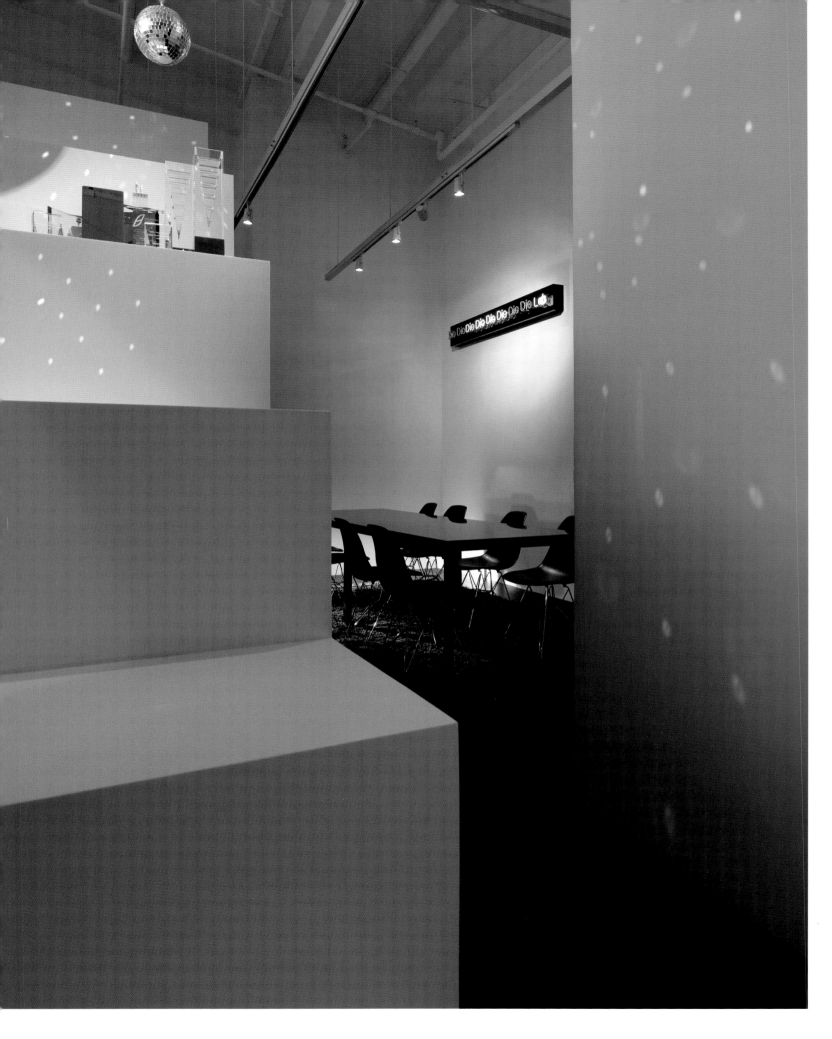

Inhabiting 6 converted shop house units at the confluence of Singapore's historic Chinatown area and the CBD, MOD's own design studio employs the same key principles which govern its approach to design - typological relevance, a disciplined material and tonal palette and an 'essentialised' concept.

In searching for a locale to base its creative endeavors, we were determined to find a building large enough to accommodate our entire office on a single floor. We feel strongly that open communication is key to innovation and creativity; as such, our space would be without hierarchy or barriers, a truly open office.

The linear series of spaces within the studio are choreographed in between the perimeter of twin datum lines -which form circulation axes spanning the entire space. Mirrored terminus points elongate these axes and become daily 'catwalks'.

Resembling a barcode, a series of mixed program are position along these catwalks and range from gallery space, meeting spaces, open plan desks, hot desk discussion zones to a library.

All new interventions are conceived as objects within the landscape of the existing space and are designed to remain visually separated. The entry Gallery space allows for constant renewal and an avenue to express ourselves without needing to reinvent the entire studio.

Our work areas are intentionally pure black and white, which allows us to tackle the wide array of creative explorations against neutral foil. The library is finished in an unfinished timber and tectonically, it provides a more relaxed environment as a counter point to the rest of the space.

Smarter manufacturing —— Top-class office integration

Smarter manufacturing —— Top-class office integration

273

Mediacom 北京办公室
Mediacom Office in Beijing

设计单位：**Mi2**
设 计 师：陈宪淳
项目地点：北京金宝大厦
建筑面积：**500 m²**
主要材料：铝板水切割、玻璃、人造石、穿孔石膏板、
　　　　　拼接块毯、人造皮革、木板烤漆
设计时间：**2011**
摄 影 师：孙翔宇

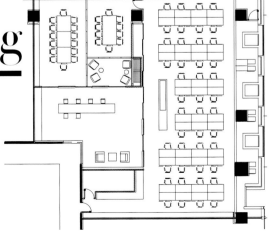

Smarter manufacturing —— Top-class office integration

备感压抑的都市人都希望找个地方释放自己,暂时逃避城市。如何才能找到更便捷的途径放松自己呢？如今,办公也可以让你逃离城市!

"逃离城市"是Mediacom全新办公空间的主题,将自然元素融合到色彩、裁剪、造型中,使人们在办公室中就能感觉到自由和释放的轻松感。

在前厅中,将繁华的城市夜景图像素化,并将这个素点图以铝板为材质制作成底墙,运用玻璃和铝板交错的方法,让人感觉自己既在城市之中,又脱离了城市的喧嚣。

在照明系统上,没有采用固有的直射照明,而是通过应用反射、单反镜等来提供光亮,以减轻灯光对办公人员的压力。在办公区中,设有植物墙和休闲炕,办公人员可在炕上阅读办公,以启发灵感。

而在会议室中,采用了蓝色、绿色和红色为主色调的块状错拼融合手法,让人在以白色为主色调的前厅和办公区之外,感受到创新和惊喜。蓝色寓意海洋,绿色寓意森林,红色寓意花朵,并且这三个大小各异的会议室可以自由组合拆分,浪漫自然之余,十分有利于办公。

贯彻一贯的设计理念,实用至上,暗藏惊喜。

Boring and full of pressure, urban people begin to seek a place to relax themselves and escape the hustle and bustle. How to get such a way to relax themselves? Today, even office can help you to escape the hustle and bustle!

"Escaping hustle and bustle" is the subject of Mediacom for brand new office space. Merging the natural elements into the color, scissoring and cutting, people can feel the freedom and sense of relaxationg in such office.

In the vestibule, put the nice urban night view on photos, and putting these spots on aluminum board as the background of wall, combined the glass and aluminum together, people will have a perfect feeling that in the city but escapt the hustle and bustle.

For the lighting system, it does not adopt direct lighting but through reflection and SLR mirror to provide light to reduce the pressure of light on office staff. In this office area, there is plant wall, leisure platform, where officers can read on to gain more inspiration.

While in the meeting room, adopted blue, green and red as the main color and collected them togerher, it has provided a sense of innovation and surprise on the vestibule and office. Blue refers to sea, green refers to forest, red to flower, and these blucks can be combined and unfolded freely, which is romantic but practical.

Upholding the usual design concept: putting practical first and providing big surprise.

Smarter manufacturing —— Top-class office integration

飞利浦创新科技园LED中心
LED Center of Philips Innovation Science and Technology Park

设 计 师：申强
项目地点：上海
建筑面积：**3700 m²**
竣工时间：**2011**

位于上海漕河泾开发区的飞利浦创新科技园LED中心，是由一个层高达8m，面积达3700m²旧厂房改造而成。改造后的LED中心以一个开放式办公区域为主体，设计师以中国传统文化为主题，在开放式办公区域一侧，巧妙地将金、木、水、火、土的风水学理念融合其中，设计5个连续的空间单体，使得整个空间不再单调而变得更为立体。由于层高较高，还设计了夹层办公区，整个空间可供200多名研究和设计人员办公。

进入飞利浦LED中心，最为突出的印象是开放式办公空间安装的圆环形灯具，数十个专为该中心定制的圆环形LED灯具，在8m高的空间中，采用悬挂式吊装，连成一片，像几十朵晶莹的花朵，颇为壮观。而圆环形灯具干净整齐的线条，显得简约，时尚而具有科技感，打破了传统灯具的印象，也为旧厂房结构改造的办公空间，增添了办公氛围，成为了整个飞利浦LED中心标志性视觉符号。整个开放办公区域开阔的空间，由于利用了LED照明和大面积房顶自然采光，丝毫没有旧厂房结构通常有的空旷感，而是给人以舒适，明亮，通透的办公氛围。

在开放式办公区域的一侧，以金、木、水、火、土为主题的会议区域，茶水间及其他休息区域，我们看到的是LED照明丰富多彩的一面。为了配合该区域的设计主题，材质应用，色彩变化，和区域的功能特点，照明设计充分运用LED光源色彩多变的特点，把该区域的照明设计为色彩可变幻的照明。和开放式办公区域明亮，通透的照明方式相比，该区域的LED照明更为活泼，多彩，营造了一种亲近，放松氛围。

通过颇具建筑美感的楼梯，来到夹层办公区域，层高近4m的夹层办公区域并不显得局促，由于房顶部分采用了大面积的自然采光，整个夹层办公区域显得十分敞亮。

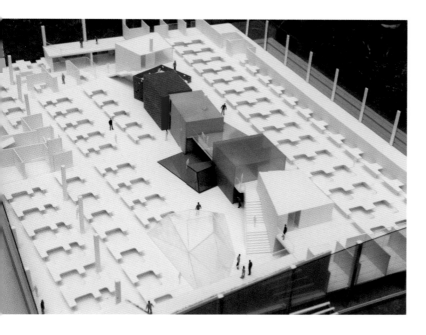

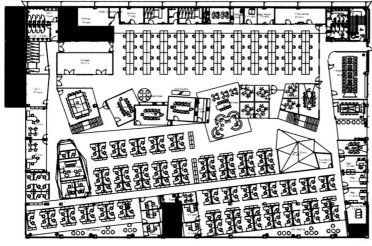

二层平面布置图

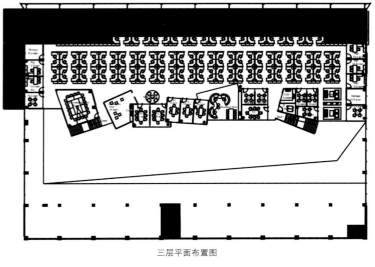

三层平面布置图

LED Center of Philips Innovation Science and Technology Park, located at Caohebang Development Zone of Shanghai, is rebuilding from an old factory with 8 meters high and 4700 square meters large. After reconstruction, this LED center takes an open office area as the main body. Upholding the traditional Traditional culture, the designers have merged senses of golden, wooden, water, fire and earth into it. They are established as five different room, allowing the whole space tridimensional but not boring. Because it is much high, the designers have placed a interlayer office area specially. The entire space is capable to hold more than 200 researchers and designers.

Entering this Philips LED center, the most attractive part is the round lighting installed in the office area. Tens of round LED lightings are hanging 8 meters high. Viewing them afar, they look like tens of crystal lotuses, charming and fascinating. The neat lines of these round lamps look more simple, fashion and full of sense of technology, which have broke the impression to traditional lamps and added much glamour to the rebuild office, becoming a landmark of vision to Philips LED Center. As utilizing the LED lightings and large scale nature light, this rebuild office has no the empty sense any more but give people a comfortable, bright and transparent space.

At one side of this open office ar-

ea, there are the meeting parts covering the rooms taking golden, wooden, water, fire and earth respectively, and the tea room as well as lounge room. The scenery we see is the most colorful one of LED. To fit the design theme of this area, the designers have adopted the colorful and changeable features of LED on selecting materials, color and functions, which is aiming to make it a changeable lighting effect. By compared with the open office's bright and the transparent manner, this area's LED lightings look more active, colorful, creating a comfortable and relax atmosphere.

Passing through the stairs that with architectural aesthetic, and entering the interlayer office, the 4 meters high office looks not small anyway. As there is large scale nature light, the whole interlayer office looks much.

Smarter manufacturing —— Top-class office integration

Smarter manufacturing —— Top-class office integration

宽银幕 大家庭
Wide Screen, Big Family

设计单位：郑州弘文建筑装饰设计有限公司
设 计 师：王政强、苏四强、仲唯伟
项目地点：郑州市郑东新区商都路5号中力
　　　　　集团大厦9层
建筑面积：600 m²
主要材料：乳胶漆、塑胶地板、生态木
设计时间：2011.09
摄 影 师：周立山

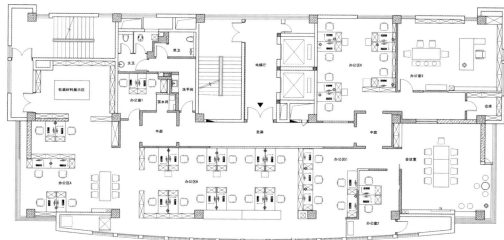

今天又是一个雾霾天,空气脏得喘不过气来,满城的汽车肆无忌惮、不知疲倦地奔跑着,成功的渴望写在每个人脸上,麻木地兴奋着,这就是我们生存的环境,欲望都市。我们处在了一个个性化、多元化的繁荣时代,但是在经济社会快速发展的背后,难掩我们心灵信仰失落的悲哀。

公司的理念、价值观和人文情感会在自己的办公空间里真情流露吗?弘文设计的办公空间不追求标新立异,刺激过载,以静雅的白色为主调,在方正平实的空间里散发出清新自然的健康活力,构建人与人之间真诚和谐的心灵,以抵抗外界的污染,像一个大家庭一样快乐成长。

新的办公室面积600m²,从9层望外,明亮宽敞视野很好。建筑本身东西狭长南北较短,因此在平面布局上分为公共区和工作区,南北向以16:9的宽银幕窗作为内外的视觉交流,既相对独立又互为联系,室内空间通过一系列的窗口形成景观互动。充满力量的直线,明快简约的白色,朴素自然的混凝土,传达了清新的特质。让设计师能静下来更加专注地投入到创意工作中。

宽银幕窗口的展现与延伸,使每个人置在其中像电影中的角色,每一天在关注与被关注中演绎着不同的感动,记录着相同的追求,见证着共同的成长。让我们不随波逐流,不卑不亢地活着!在这里,空间的设计和设计的空间有了新的意义……

It is a foggy day again. Dirty air makes people cannot breathe well. Cars are running between streets among the whole city freely and vigorously. The desire to success is showing on each face, coldly but exciting. That's our living environment, sex and the city. We are living in a personalized and diversified booming times. However, behind the rapid economic and social development, nothing can conceal the sorrow of losing our spiritual beliefs.

Can the company's concept, value and human emotion be expressed among our office? Without pursuing of unique and overstate, Hongwen Design takes white color as the tone. Among the square space, it is showing a fresh and health sense, connecting people's sincere and harmony hearts to resist the outside hustle and bustle, growing like a big family with happiness.

The new office area is 600 square meters. Viewing from the 9th floor, it is great with brightness and spacious. The building is long and narrow in east-west direction and short in south-north direction. Therefore, there is divided into public area and office

Smarter manufacturing —— Top-class office integration

area. The 16:9 wide screen window in the south-north direction is a platform as the visual communication between internal and external. The internal space is exchanged with the series of window landscape. The vigorous straight line, simple and bright white color, natural cement is outpouring a sense of fresh and clean. The designers can pay more attention on creation.

The expression and extension of wide screen window allow each person to play a role like in a film. There is illustrating different moving and touching among the focusing and concern, recording the same pursuit, witnesses the growth. Therefore, living with our own belief is a great choice. Here, the design and space have been endowed with new significance…

Smarter manufacturing —— Top-class office integration

Vital Déco 办公室
Vital Déco Office

设计单位：**Frans Chan Design International**
设 计 师：陈方晓
项目地点：福建厦门
建筑面积：**1100 m²**
主要材料：水泥、亚麻、锈钢板、深黑氟碳漆、银灰色金属漆、深色大理石
摄 影 师：陈方晓

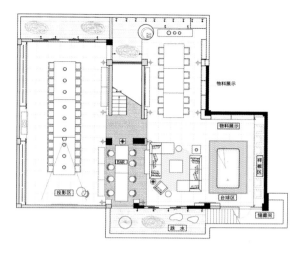
负一层平面布置图

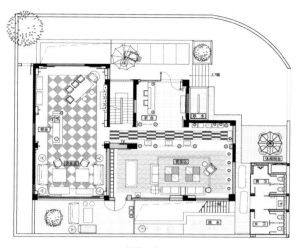
一层平面布置图

讲设计不如讲故事，做造型不如做光影。设计师运用光影来塑造空间的叙事性和戏剧效果。空间造型不再是单一的审美需求和使用功能，设计师赋予其文化内涵。

在会所墙上有这么几段话："楼梯、放慢脚步，让灵魂跟上。""无就是有，有就是无。""越危险的地方就越安全。"

空间也应是时间的产物，它也有童年，青春……让墙面的锈迹记显青春的痕迹。

左手，右手，亦或转身，爱情就在那，转瞬即失，机会亦然。选择是人生的首要之事，只要选择了就需为所选执着。

Talking about design is inferior to story telling, doing styling is inferior to shadow. The designers use light and shadow to shape the narrative and dramatic effect of the space. Space modeling is no longer a single aesthetic needs and use, the designers give its cultural connotations. Take it easy, Let your soul catch up. Form does not differ from the void, and the void does not differ from form. More dangerous place is safer.

Space is produced by time, at the same time come with childhood, youth……Let's record youth via rust of surface of wall……

His left hand, right hand, also or turn around, love is in that, soon that is lost, the opportunity vice versa. Life is the first choice of, as long as the choiceof the needs for the selected persistent.

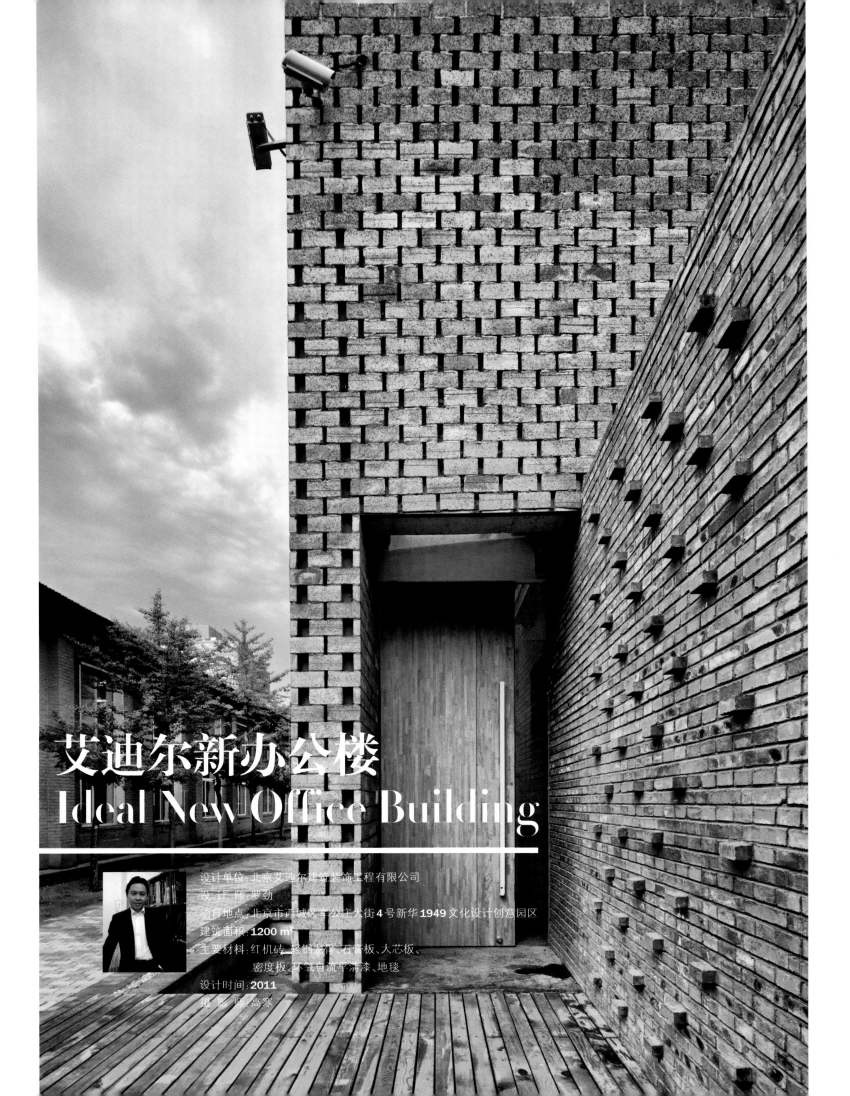

艾迪尔新办公楼
Ideal New Office Building

设计单位：北京艾迪尔建筑装饰工程有限公司
设 计 师：罗劲
项目地点：北京市西城区车公庄大街4号新华1949文化设计创意园区
建筑面积：1200 m²
主要材料：红机砖、轻钢龙骨、石膏板、大芯板、
　　　　　密度板、环氧自流平清漆、地毯
设计时间：2011
摄 影 师：高寒

这是一栋老仓库改建项目。营建风格独特且具有包容性和多样性的商务空间,打造能够鼓舞和激励员工的现代高效办公场所,是我们此次改造的重要目标和旨意。

我们局部下挖了仓库的内部地面,加建出错落的层次空间,提高了建筑的使用效率。增建的夹层空间从东向西展开并将室内以多种形式围合起来。局部的矮墙、石条、台架、水景、悬梯、廊桥等作为过渡构件,成为连接各个部门单元的重要组成部分,共同营建了人看人、空间套空间、围合包围合的丰富办公组团环境。我们在原有建筑的四面都适度进行了加建,营造出门斗、外廊、健身阳光房、员工餐厅以及多功能厅等空间,这些增建空间与院墙及周边建筑之间形成不同尺度和感受的围合形态,使得原本呆板的建筑外部表情丰富多彩。根据原有建筑的空间特点和结构形式,我们尽量考虑在室内全方位引入自然光,增加了顶部的天光照明,提高了室内的光照及感官效果。

我们成功申报了LEED金奖认证。通过设计与技术的结合,我们试图在节水节能节材,资源的循环利用等各方面认真考虑;在设计到施工的各个环节上细微处理,真正打造一个较为全面的绿色环保办公环境。

It is a rebuild project for an old warehouse, which is aiming to create an unique but diversity commercial space, making a modern and efficient office space to encourage and inspire staff, as well as the most important goal and message of this project.

We have dug the inner floor of the warehouse so as to enhance the scattered-level space and improve the efficiency in the using of the building. The added mezzanine space spreads from east to west and surrounds the inner door in various ways. The partial low wall, stone, bench, water scenery, hanging ladder, bridges, etc, are all the transitional elements, which have become the major parts of connecting the departments. They have built a wonderful office groups environment that merge people, space in space and room in room together. Based on the former warehouse, we have created the door socket, verandahs, fitness room, staff

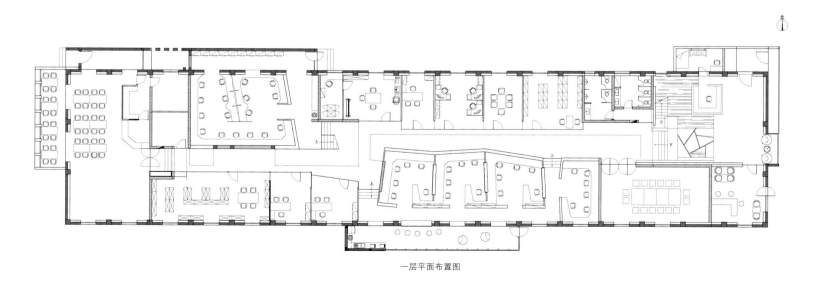

一层平面布置图

restaurant and multi-function rooms and other space, which have formed different sizes and feelings for the space and the surrounding walls and buildings, giving more colorful appearances on the former boring building. According to the former space features and construction structure, we are trying to introduce
the natural light and add the skylight on the top to improve the indoor light and sensory effect.

We have applied the LEED Gold Certificate successful. Through combining the design and technology, we are trying to take the energy saving and material saving, recycling of resources into consideration. We have given much subtle treatment on different aspects of the construction, which has created a much more comprehensive green office environment truly.

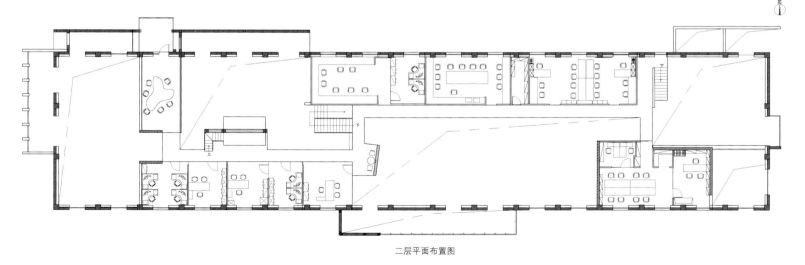

二层平面布置图

Smarter manufacturing —— Top-class office integration

真工设计黎明事务所办公室
Office of Z-WORK Design Associate

设计单位：真工设计工程股份有限公司
　　　　　/Z-WORK Design Associate
设 计 师：程绍正韬
项目地点：台湾台中
建筑面积：北栋 **165m²**、南栋 **99m²**
主要材料：油漆、水泥、绿色植物
设计时间：**2010**
摄 影 师：李国民

真工设计黎明事务所正位于城市新旧发展纹理的交接面上,我们在处理外部建筑与内部空间设计的同时,首先思考的是如何利用都市设计上模糊边界概念(Boundaryless)来对应既成文化涵构已经存在已久的意义。于是,从领域的模糊性(Fusion Field)议题;到中介性(inbetween space)的空间内涵,便发生在像台中市干城街黎明社区这类50年历史的公家眷舍基地上。对目前的都市新区扩张现况来说,那正是一个所谓旧社区与新城区交界之接壤区域,也是一个珍贵的地方性历史人文永续发展与现代性资本地产短利操作之间,互相冲撞的尴尬领域。

黎明社区历经长时间人文与生态平衡之后,已经成为极宜人居的人性社区了;它不仅没有现代大都会的那种拥挤、躁动、不安等特质,还常常是现代都市居民做为定居场所的珍贵首选;合宜的街廓布局中散布着许多低矮的巷弄,人安定地生活于鸟语花香之间,人文与生态平衡后的幸福画面,就这样自然地间杂在这种新旧域区交界的都会绿带中。这个人文多元性与生态多样性的新中介空间,虽然区隔了都市新区与传统生活的旧世界,但而却成为当代水泥丛林化现代都市中,一种保护人与生态共生的间隔区段。如果这类已逐渐共生化的社

区,在未来台湾都市发展中可以被保留,甚至其人文与生态发展之进程可以被借鉴为和谐城市之发展法则之一,那么,我们的都市更新策略,便可以在这种现代城市中新的人为环境开发与自然共生的机制系统中找到平衡,也找到一种幸福。

做为一个环境发展人与自然共生理念的实践者,强化环境、弱化建筑的美学概念倡议者及设计工作与凡俗生活、生命修行合而为一的自我革命者,我们在黎明社区边壤的事务所采取的就是一个这样低调且反省的态度,事实上一般设计者会利用设计造型上的突出来强调自我,但也会一定程度对立了社区人文与自然之间的平衡关系。因此,这个时候倒不如静下心来、慢下表现步调来,用一种向内看的禅者之心,澄怀味象中让自己的设计消融,臣服在这片绿色的凡俗大海中。

真工的黎明事务所,就是立基于这样的一个环境创作条件:城市文化与生态过渡地带的敏感场所性质;另外,这个真工的上位计划事务所,它主要的目的是为了海峡两岸的设计分工整合的目的而存在;这当然也是一种经济事务上的中介过渡性格,在繁忙与庞杂中统合与施令就是它的工作特性。

The Office of Z-WORK Design Associate in Liming Community is exactly located on the interface between the new and old developments of a city. Therefore, in treatment of the designs of both the interior and exterior of the building, the first priority was given to application of the boundarylessness, a concept used in urban design, to the long-lasting importance of the existing cultural compositions. Hence, the discussions about fusion fields and the connotations of the in-between spaces arose in relation to the historical premises like the Liming Community, a 50-year-old housing existence for the public located in Gancheng Street, Taichung. For the current expansions of the new areas in a city, the community is exactly the area where old buildings join the new ones and an embarrassing field where collisions between the sustainable development of a precious local historical heritage and operations of modern capitalists for the short-term benefits occur.

After the long balance between the civilization and the ecology, Liming Community has become a very habitable and humane society; without high population and restlessness, which are features of a modern metropolis, it is often a valuable alternative when people consider a place for themselves to live and work peacefully and contently. particularly, a lot of alleys scatter in the streets of moderate sizes and people live tranquilly in birds' chirps and scents of flowers and the satisfactory pictures brought by the balance between civilization and ecology can be seen in the oasis where old and new developments join. Although the new inbetween space for diversities in civilization and ecology separate the new urban areas from the traditional old world, it serves as a partition to preserve the coexistence of people with the environment in a modern metropolis characterized by ferroconcrete jungles; if the coexistence found in this kind of communities can be preserved for the development of Taiwan in future and further the process of development of the civilization and the ecology can be introduced as the principle for a city to develop harmoniously, then, we will be more strategic in development of our city and we may find a balance or a content in the system where artificial environment development of a city co-exist harmoniously with the nature

As the practicer of the concept

that man and the nature should co-exist harmoniously in development of the environment, advocate of the aesthetic concept that the environment should be given more importance than the construction and the self-conscious revolutionary who fuse the design job with worldly life and austerity practicing, we took a low-profile and introspection attitude toward the design of the office on the border of the Liming Community. In fact, the way that most architects take to make himself impressive by making his designs distinct has made his products incompatible with the environment to some degree. Hence, it is better to settle yourself down to make your designs less obtrusive and fuse your products with the environment nearby and subject to it with a buddhistic attitude

The Office of Z-WORK Design Associate in Liming Community is designed to fit itself to such an environment as a sensitive land where city culture flow to the ecology; on the other hand, as the office of Z-WORK for plans in authority, the office is designed to coordinate the design jobs between the architects across the straits, which can also be seen a transitional institution, the functions of which are busy and numerous for coordinations and decision makings.

Smarter manufacturing —— Top-class office integration

艺谷（北京）公司新办公室
The New Beijing Office Eegoo

设计单位：dEEP Architects
设 计 师：李道德
项目地点：北京
建筑面积：**2300 m²**

Smarter manufacturing —— Top-class office integration

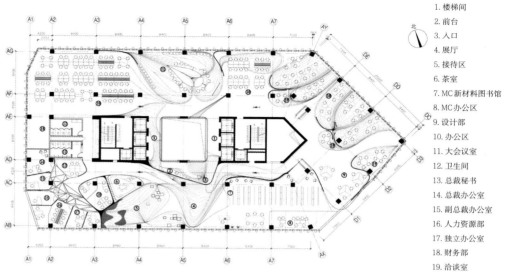

1. 楼梯间
2. 前台
3. 入口
4. 展厅
5. 接待区
6. 茶室
7. MC新材料图书馆
8. MC办公区
9. 设计部
10. 办公区
11. 大会议室
12. 卫生间
13. 总裁秘书
14. 总裁办公室
15. 副总裁办公室
16. 人力资源部
17. 独立办公室
18. 财务部
19. 洽谈室

传统办公室布局和构形多采用立方体，艺谷（北京）公司新办公大厅的设计则与此截然相反。有些建筑师认为建筑物的形状应服从其功能。有些建筑师则反其道而行之，就像你在"艺谷"所能感受到的淙淙流水般的线条和峰回路转的结构，而这些都应归功于一个接一个的细胞结构。这些细胞结构彼此类似而又浑然天成，功能齐备而又变化多端，可谓另类。北京艺谷办公大厅是民主讨论与集中统一和谐相处的场所，在这里你的思想可以自由流淌，由细胞到细胞，由私有变公有。

该办公大厅采用节点布局，各细胞的面积与用途皆无定式，忽而是隐秘的办公室，忽而又是图书馆，走着走着又出现了一个咖啡厅。曲径围绕着办公大厅的核心在流淌，在某些地方的曲径又豁

317

然开朗,变成一个个功能单元。在咖啡厅、图书馆等较大的公共区域曲径最粗,而在相对较小的办公室附近曲径又变细。

表面的处理模拟了流水的形态,表现为柔软材料到硬质材料平滑过渡。咨询台好似一圈圈涟漪,而有些小办公室周围的冠盖其实是一些表面磨光的功能单元。与此类似,各个表面功能也表现平滑过渡,墙面流淌为地面,而天花板则流淌为细胞膜。

小细胞可以彼此融合变大,成为公私两用的空间。中心会议室由三间较小的细胞单元构成,这些小细胞单元可以扩展成为一间较大的细胞单元。每个单元因一些提示性的界限而泾渭分明,一部分界限是固定的墙体或框架,而另外一些界限则由一些柔性材料构成,如窗帘,这样小空间就可以彼此融合成为大的公用空间。按照这种思路,大会议桌也可以拆分为小会议桌供小空间的人使用。

空间忽而收缩,忽而扩张,了无痕迹,这正应了艺谷公司现阶段的动态变化。当办公室的员工各司其职时,墙面和窗帘为他们划分出了小的、独立的空间或小室,而当整个办公室需要共享各自的思维时,这些墙体又消失了,成了更大的细胞,供员工畅所欲言,超越彼此的界限。

Smarter manufacturing —— Top-class office integration

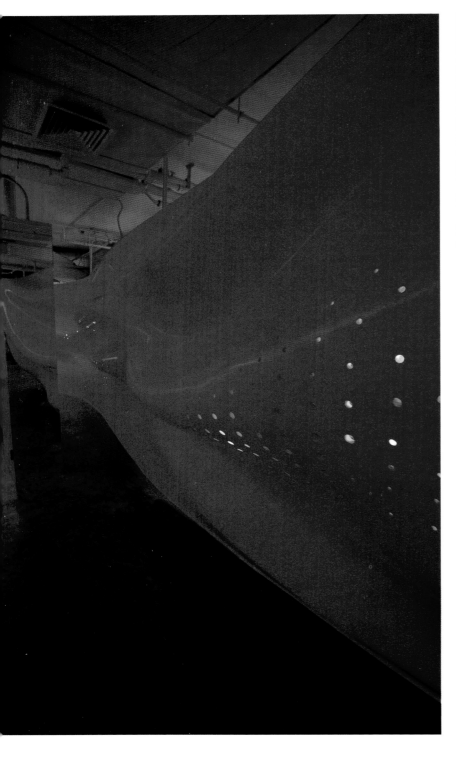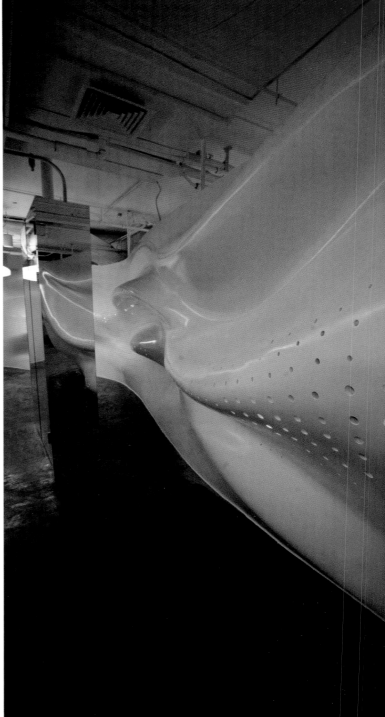

Smarter manufacturing —— *Top-class office integration*

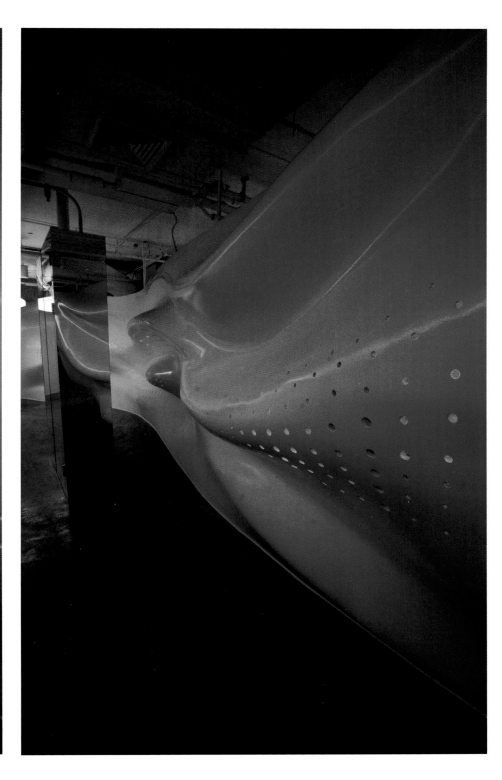

The new Beijing office for eegoo is the antithesis to the office typology where traditionally the cubical dominates the organization and shape of the program. As some will argue, form follows function. Conversely, what can be experienced is a flow of form and circulation generated by a cellular sequence. The cell structure generates a congruent office environment while enlisting a variable of juxtaposing functions not typically found in the typology. The Beijing eegoo office is a space for open discussions and decision making to be accord; an office where ideas can flow freely from the private to the public, from cell to cell.

A nodal organization sequence is given to the program, in which generated cells range in size and use, from private offices to a library and a cafe. The circulation meanders around the building's core and swells in size to accommodate surrounding functions. The path is largest near the larger public cells of the café and materials library, and narrower around the smaller private offices.

The treatment of surfaces imitates the flowing experience, expressed as a gradient of soft to hard materials. The reception desk features liquid-like ripples, where as a canopy near the small offices is made of faceted cells. Likewise, the

function of each surface expresses a gradient where walls flow to become either the ground or ceiling creating the cells' membrane.

The smaller cells have the ability to fuse and create larger public private spaces. The central meeting room is organized by three smaller cells which can be expanded into one large cell. Each cell is articulated with suggestive boundaries, one half a permanent wall or frame, the other half a pliable material like a curtain to allow for the spaces to fuse together and create general assemblages. Mimicking this concept, the office table can be separated into three smaller meeting tables to be used in the new spaces.

The space effortlessly contracts and expands to emulate the dynamics of the eegoo office in the present moment. As the office is working as individuals, walls and curtains divide the space into smaller, intimate rooms or cells. When the office is sharing ideas, the walls are removed, creating larger cells where conversations will cross boundaries.

Smarter manufacturing —— Top-class office integration

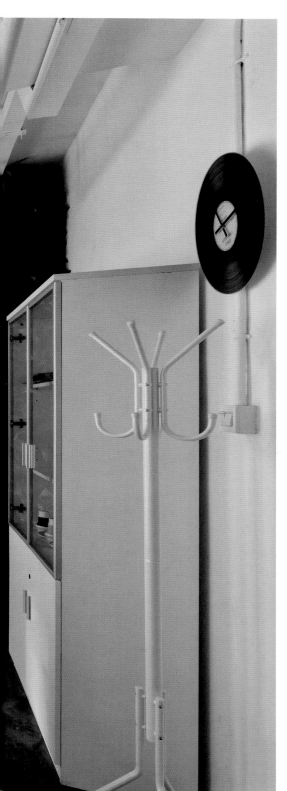

Smarter manufacturing —— Top-class office integration

图书在版编目(CIP)数据

智造：顶级办公设计集成：汉英对照 /《智造：顶级办公设计集成》编委会编. -- 北京：中国林业出版社, 2012.9
ISBN 978-7-5038-6737-8

Ⅰ.①智… Ⅱ.①智… Ⅲ.①办公室—室内装饰设计—作品集—世界 Ⅳ.①TU243

中国版本图书馆CIP数据核字(2012)第213343号

【智造：顶级办公设计集成】编委会

本册顾问：陈宪淳

编　委：Dariel Studio　　Marcelo Joulia　　林伟明　　邵唯宴
　　　　黄士华　　　　　吴联旭　　　　　罗　劲　　汤建松

编写成员：孔新民　贾　刚　高囡囡　王　超　刘　杰　孙　宇　李一茹
　　　　　姜　琳　赵天一　李成伟　王琳琳　王为伟　李金斤　王明明
　　　　　石　芳　王　博　徐　健　齐　碧　阮秋艳　王　野　刘　洋
　　　　　陈圆圆　陈科深　吴宜泽　沈洪丹　韩秀夫　牟婷婷　朱　博
　　　　　宁　爽　刘　帅　宋晓威　陈书争　高晓欣　包玲利　郭海娇
　　　　　牛晓霆　张文媛　陆　露　何海珍　刘　婕　夏　雪　王　娟
　　　　　黄　丽　程艳平　高丽媚　汪三红　肖　聪　张雨来　韩培培

采访整理：柳素荣
责任编辑：纪　亮　李丝丝

中国林业出版社 · 建筑与家居出版中心

出版：中国林业出版社　（100009 北京西城区德内大街刘海胡同7号）
网址：http://lycb.forestry.gov.cn
E-mail：cfphz@public.bta.net.cn
电话：（010）8322 3051
发行：新华书店
印刷：北京利丰雅高长城印刷有限公司
版次：2012年9月第1版
印次：2012年9月第1次
开本：230mm×300mm　1/16
印张：21
字数：180千字
定价：320.00元

购买本书凭密码赠送高清电子书
密码索取方式
QQ：179867195，E-mail：frontlinebook@126.com

法律顾问：北京华泰律师事务所 王海东律师 邮箱：prewang@163.com